Tactics of Being An Agripreuner-Learning the Rope

Tactics of Being An Agripreuner-Learning the Rope

Dr. G.S. Sreedaya
Assistant Professor (Senior Scale)
Kerala Agricultural University
Department of Agricultural Extension
College of Agriculture, Vellayani,
Thiruvananthapuram-695522, Kerala

Dr. A. Anil Kumar
Faculty Dean
Kerala Agricultural University
College of Agriculture, Vellayani,
Thiruvananthapuram-695522, Kerala

&

Dr. B. Seema
Professor & Head, ADE (SZ)
Kerala Agricultural University
Department of Agricultural Extension
College of Agriculture, Vellayani,
Thiruvananthapuram-695522, Kerala

SSPH

SATISH SERIAL PUBLISHING HOUSE
403, Express Tower, Commercial Complex
Azadpur, Delhi-110033 (India)
Phone : 011-27672852, Fax : 91-11-27672046
e-mail : info@satishserial.com, hkjain1975@yahoo.com
Website : www.satishserial.com

Published by :

SATISH SERIAL PUBLISHING HOUSE

403, Express Tower, Commercial Complex, Azadpur, Delhi-110033 (INDIA)
Phone : 011-27672852 Fax : 91-11-27672046
E-mail : info@satishserial.com, hkjain1975@yahoo.com

ISBN 978-93-88020-48-0

Composed, Designed & Printed in India

Foreword

A paradigm shift from Agriculture to agri-business is an essential pathway to revitalize Indian agricultural scenario and to make agriculture more attractive and a profitable venture. Agripreneurship have the potential to contribute to a range of social and economic development, such as employment generation, income generation, poverty reduction and improvements in nutrition, health and overall food security in the national economy.

Agripreneurship is not only an opportunity but also a necessity for improving the production and profitability in agricultural sector. To be a successful agripreneur, a farmer needs to be innovative, profit-oriented and risk taker. The Entrepreneurial development process refers to inculcating the entrepreneurial skills which include soft skills, technical skills, financial skill and managerial skill into a person to build the entrepreneurial approach.

The sincere efforts taken for publication of the book "**Tactics of Being An Agripreneur: Learning the Rope**" contributed by the scientists and research scholars of Kerala Agricultural University(KAU), Kerala Start up Mission, Central Tuber Crop Research Institute (CTCRI), Institute of Management in Government(IMG) and Kerala Veterinary and Animal Science University (KVASU) are commendable and the contents are designed to impart an awareness regarding various dimensions of agripreneurship. I congratulate the editors and the contributors for their sincere efforts in bringing out this publication.

Let us hope this publication would be highly useful, in one way or the other, for students, researchers, academicians, extension workers, policy-makers, planners, officials in development institutions/agencies and above all the agripreneurs.

Dr. R. Chandra Babu
Vice Chancellor
Kerala Agricultural University, Thrissur

Foreword

Agripreneurship is the key to rejuvenate agricultural sector in India. It is not only an opportunity but also a necessity for improving the production and profitability of agriculture. The immense potential of Agripreneurship and agri-business would surely be a catalyst for employment generation, income generation, poverty reduction and improvement in nutrition, health and overall food security.

Sincere efforts taken for publication of the book "**Tactics of Being An Agripreneur: Learning the Rope**" with contributions from the scientists and research scholars of Kerala Agricultural University (KAU), Kerala Startup Mission, Central Tuber Crop Research Institute (CTCRI), Institute of Management in Government (IMG) and Kerala Veterinary and Animal Science University (KVASU) are commendable. This book, I believe, would impart necessary awareness on the entrepreneurial and technological skills to start a venture in agriculture.

I congratulate the editors and the contributors for their sincere efforts in bringing out this important publication.

Let us hope that this book would immensely empower prospective entrepreneurs with adequate knowledge on agripreneurship and skills in managing enterprises scientifically.

Dr. Jiju P Alex
Director of Extension
Kerala Agricultural University
Thrissur

Preface

Agriculture and allied sectors are considered to be the mainstay of the Indian economy. It forms the important source of raw material and demand for many industrial products. A shift from agriculture to agri-business is an essential pathway to revitalize Indian agriculture and to make more attractive and profitable venture. Agripreneurship is the profitable marriage of agriculture and entrepreneurship more plainly turning the farm into a business. Agripreneur is defined as entrepreneur whose main business is agriculture or agriculture-related activities. Agripreneurship can be used as best medicine for the solution of various problems viz, reduce the burden of agriculture, generate employment opportunities for rural youth control migration from rural to urban areas, increase national income, support industries development in rural areas, reduces the pressure on urban cities, etc. Agripreneurship also play in inducing productivity gains by smallholder farmers and integrating them into local, national and international markets, reduce food costs, reduce uncertainties and improving the diets of the rural and urban poor in the country and to generate growth, increase and diversify income by providing entrepreneurial opportunities in both, rural and urban areas. Agripreneurship have wide range scope in various sectors like agriculture, horticulture, animal husbandry, fishery and others. Group entrepreneurship is particularly attractive among those farmers who would not be able to start an entrepreneurial business on their own. Establishment of collection centers with warehousing facilities and e-marketing facility in rural areas and establishment of cold supply chain will provide employment opportunity to the youth. Creation of agri-clinics and agri-business centers managed by agri-graduates so as to provide consultancy services to the farming community in rural areas an attractive venture. Finishing School for V.H.S.E (Agri) Certificate Holder and various training institutes also play an important role in entrepreneurship development. Government programmes like 'Start-Up India', National Skill Development Mission, and 'Rural Entrepreneurship and Awareness Development Yojana' (READY) programme envisaged in the Twelfth Five Year Plan aims at entrepreneurship development among the rural youth.

Keeping these things, the book entitled **"Tactics of Being An Agripreneur-Learning the Rope"** was visualized and we first thought about writing each chapter ourselves, but we quickly realized that both, technicalities and techniques are essential to run an enterprise and therefore, requested the contribution of experts at various fields related to agriculture. The book is divided into five sessions viz, Entrepreneurship Development Process, Crop Production, Crop Management, Crop Improvement, Agriculture and Allied Ativities and Livestock Production Management.

We strongly believe that this robust rendering would be the most informative and powerful way to ensure that our readers could understand the areas of agripreneurship. Of course, we made sure that authors used a similar chapter structure for easier reading, but we relished in the fact that each chapter also has its own voice. Because of this, the book can be read cover to cover, or you can pick any chapter and read it on its own. Hopefully, you will find that the more you read, the more you will be inspired, educated and empowered to help move yourself to an agripreneur.

We profusely thank Dr. R. Chandra Babu Vice Chancellor, Kerala Agricultural University, Thrissur for his kind support in completing this venture. We also express our sincere gratitude to Dr. Jiju P. Alex, Director of Extension, Kerala Agricultural University, Thrissur for his profuse support in completing the venture. As editors of this book, we would like to thank all the authors for their efforts and co-operation in bringing out this book well in time. We would like to give special mention to Dr. Saji Gopinath, CEO Startup mission, Kerala who has found valuable time in contributing a chapter on startups. We thank all the teachers, students and staff of Deptt. of Agri. Extension College of Agricultue, Vellayani, who have helped us to complete this endeavour. We firmly believe that this publication will be highly useful, in one way or the other, for students, researchers, academicians, extension workers, policy-makers, planners, officials in development institutions/agencies, producers and students.

Editors

List of Contributors

Dr. A. Anil Kumar
Faculty Dean
Kerala Agricultural University
College of Agricuture, Vellayani,
Thiruvananthapuram-695 522, Kerala

Dr. Saji Gopinath
CEO Kerala Startup Mission
G3B,Thejaswini,Technopark,
Thiruvananthapuram-695 522, Kerala

Dr. P. Sethuraman Shivakumar D.
Central Tuber Crops Research
Institute, Sreekaaryam,
Thiruvananthapuram-695 522, Kerala

Dr. A.S. Anilkumar
Associate Director of Research(SZ)
Kerala Agricultural University
College of Agriculture, Vellayani,
Thiruvananthapuram-695 522,
Kerala

Dr. Suraj P.T.
Assistant Professor (Sr. Scale)
Department of LPM
College of Veterinary and Animal
sciences, Mannuthy, Thrissur

Dr. Sajeena S.
Assistant Professor (Sr. Scale)
KVK Malappuram
KCAET Campus, Tavanur
Malapuram

Dr. N. Kishore Kumar
Professor & Head
Training Service Scheme
College of Agriculture, Vellayani,
Thiruvananthapuram-695 522,
Kerala

Dr. B. Seema
Professor & Head
Agrl. Extension & Associate
Director of Extension (SZ)
College of Agriculture, Vellayani,
Thiruvananthapuram-695 522, Kerala

Dr. Hajilal M.S.
Professor and Head
Department of Agricultural
Engineering, College of Agriculture,
Vellayani, Thiruvananthapuram-
695 522, Kerala

Dr. T. Sajitharani
Professor & Head, Instructional Farm,
College of Agriculture, Vellayani,
Thiruvananthapuram-695 522, Kerala

Dr. A.K. Sherif
Nodal Officer, Training Service
Scheme, College of Agriculture,
Thiruvananthapuram-695 522,
Kerala

Dr. Xavier K. Jacob
Professor & Head
Department of Agricultural
Engineering, College of Agriculture,
Vellayani, Thiruvananthapuram-
695 522, Kerala

Dr. Roy Stephen
Professor, Department of Plant
Physiology, College of Agriculture,
Vellayani, Thiruvananthapuram-
695 522, Kerala

Dr. K.S. Meenakumari
Professor & Head
Department of Agricultural
Microbiology, College of Agriculture,
Vellayani, Thiruvananthapuram-
695 522, Kerala

Dr. Mini C.
Professor & Head, Department of Post-Harvest Technology
College of Agriculture, Vellayani
Thiruvananthapuram-695 522, Kerala

Dr. Suma Divakar
Professor, Department of Community Science, College of Agriculture, Vellayani, Thiruvananthapuram-695 522, Kerala

Dr. Geetha D.
Professor, Instructional Farm, College of Agriculture, Vellayani, Thiruvananthapuram-695 522, Kerala

Dr. R.V. Manju
Professor
Department of Plant Physiology, College of Agriculture, Vellayani, Thiruvananthapuram-695 522, Kerala

Dr. Sharu S.R.
Assistant Professor
Department of Agronomy
College of Agriculture, Vellayani, Thiruvananthapuram-695 522, Kerala

Dr. V.B. Padmanabhan
Professor and Head (Retd.), Department of Agricultural Extension, College of Agricuture, Vellayani, Thiruvananthapuram-695 522, Kerala

Dr. Swapna Alex
Professor & Head
Department of Plant Biotechnology, College of Agriculture, Vellayani, Thiruvananthapuram-695 522, Kerala

Dr. Deepa S. Nair
Assistant Professor (Sr. Scale) & Head
Department of Plantation Crops and Spices, College of Agriculture, Vellayani, Thiruvananthapuram-695 522, Kerala

Dr. Sreekala G.S.
Assistant Professor (Sr. Scale)
Department of Plantation Crops andSpices, College of Agriculture, Vellayani, Thiruvananthapuram-695 522, Kerala

Dr. Usha C. Thomas
Assistant Professor (Sr. Scale)
AICRP on Forage Crops and Utilisation, College of Agriculture, Vellayani, Thiruvananthapuram-695 522, Kerala

Dr. Ameena M.
Assistant Professor (Sr. Scale)
Department of Agronomy, College of Agriculture, Vellayani, Thiruvananthapuram-695 522, Kerala

Sri. T. Paul Lazarus
Assistant Professor (Sr. Scale)
Department of Agricultural Economics, College of Agriculture, Vellayani
Thiruvananthapuram-695 522, Kerala

Dr. Ambily Paul
Assistant Professor (Sr. Scale)
Department of Agricultural Entomology, College of Agricuture, Vellayani,
Thiruvananthapuram-695 522, Kerala

Dr. Beena R.
Assistant Professor (Sr. Scale)
Department of Plant Physiology, College of Agriculture, Vellayani, Thiruvananthapuram-695 522, Kerala

Dr. Aparna B.
Assistant Professor (Sr. Scale)
Department of Soil Science and Agricultural Chemistry,
College of Agriculture, Vellayani, Thiruvananthapuram-695 522, Kerala

Dr. P.R. Geethalekshmi
Assistant Professor (Sr. Scale)
Department of Post-Harvest Technology, College of Agriculture, Vellayani, Thiruvananthapuram-695 522, Kerala

Dr. Amritha V.S.
Assistant Professor (Sr. Scale)
Department of Agricultural Entomology, College of Agriculture, Vellayani, Thiruvananthapuram-695 522, Kerala

Dr. G.S. Sreedaya
Assistant Professor (Sr. Scale)
Department of Agrl. Extension
College of Agriculture, Vellayani, Thiruvananthapuram-695 522, Kerala

Ms. Dhanusha Balakrishan
Ph.D Scholar
Department of Agricultural Extension, College of Agriculture, Vellayani, Thiruvananthapuram-695 522, Kerala

Ms. Preethu K. Paul
Subject Matter Specialist
ICAR Krishi Vigyan Kendra
Santhanpara 685 619, Idukki, Kerala

Ms. Greeshma Udayan
Research Scholar, Department of Agricultural Extension,
College of Agricuture, Vellayani, Thiruvananthapuram-695 522, Kerala

Ms. Kavya S.R.
Research Scholar
Department of Agricultural Extension, College of Agricuture, Vellayani, Thiruvananthapuram-695 522, Kerala

Mr. Hari Sankar S.S.
Research Scholar
Department of Agricultural Entomology, College of Agricuture, Vellayani, Thiruvananthapuram-695 522, Kerala

Ms. Gayathri P. Bose
Research Scholar, Department of Agricultural Entomology,
College of Agriculture, Vellayani, Thiruvananthapuram-695 522, Kerala

Contents

I. Entrepreneurship Development Process

Chapter 1

Fundamentals of Entrepreneurship

Dr. G.S. Sreedaya (Asst. Professor Sr.Scale), **Suma Rose Sundaran, Revathy Chandran** and **Reshma Murugan** (Research Scholars)

Department of Agricultural Extension, Kerala Agricultural University, College of Agriculture Vellayani, Thiruvananthapuram-695 522, Kerala

Introduction

"I'm convinced that about half of what separates successful entrepreneurs from the nonsuccessful ones is pure perseverance."

Steve Jobs, Former Chairman and CEO of Apple

Entrepreneurship is the name given to the factor of production which performs the functions of Enterprise. Land, Labour, Capital, Organisation and Enterprise are the five factors, which are thought to be the basis of all the production activities. Entrepreneurship in a broader sense can be considered as a process of action undertaken by an entrepreneur (Person) to establish his enterprise. Entrepreneur is a person who sets up a business or businesses, taking on financial risks in the hope of profit. In other words, Entrepreneur is the person who introduces new things in the economy and is considered as the business leader and not as simple owner of capital. The term "entrepreneur" is a French word derived from the verb "entreprendre", which means to do or to undertake. It can be divided into two parts, "entre", meaning "between", and "preneur" meaning "taker". Literally, then, an *entrepreneur* is a "between-taker", or "go-between.

Pioneers of Entrepreneurship

The concept of entrepreneurship is mainly understood from the works of pioneers like Richard Cantillon, Jean Baptiste Say, Joseph Alois Schumpeter, etc. Richard Cantillon could be better described as venture capitalist; a venture looking for investment opportunities with better than average yields. His perspective as an investor-oriented around the element of 'risk' and was a core aspect of how he viewed entrepreneurial projects and defined what he considered to be an entrepreneur. Richard Cantillon described the entrepreneur as a person who purchases a raw material at a known price in order to sell it at an unknown price (Cantillon, 1755). In Cantillon's definition, an entrepreneur's role lies between that of two or more other actors. He or she is an intermediary (or go-between) who instigates a transformation.

Jean-Baptiste Say who originally belonged to an entrepreneurial family tried to convince the importance of entrepreneurship in economic development. He regarded entrepreneurs as being people who could do new things, people who could do more with less, and people who would obtain more by doing something in a new or different way (Say, 1815; 1996) and thus, added an element of innovation.

Joseph Alois Schumpeter postulates pointed out that the essence of entrepreneurship lies in the perception and exploitation of new opportunities (Schumpeter, 1928). He associated innovation by entrepreneurs with five elements:

1. The introduction of a new good;
2. The introduction of a new method of production;
3. The opening of a new market;
4. The conquest of a new source of supply of raw material;
5. The carrying out of the new organization of any industry.

Meaning and Definition of Entrepreneurship

An entrepreneurship is a combination of entrepreneur, enterprise and environment.

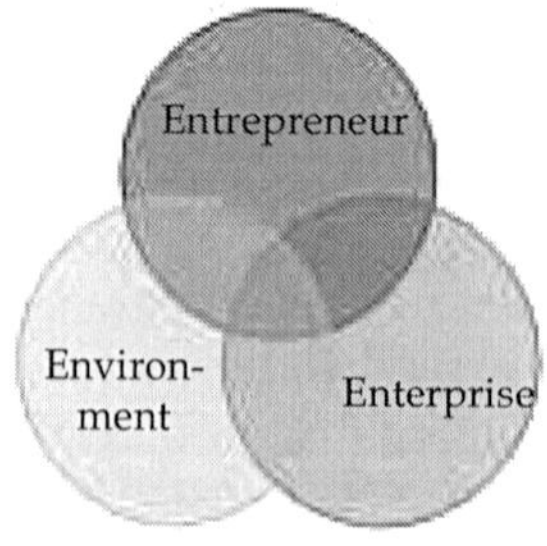

Fig. 1. Components of entrepreneurship

M.M.P. Akhouri, 1979 describes entrepreneur as a character who combines innovativeness, readiness to take risk, sensing opportunities, identifying and mobilizing potential resources, concerns for excellence and who is persistent in achieving the goal. An enterprise is a continuous process of transforming any activity or investment into an innovative venture generating relentlessly jobs and ideas, resource and value. The third factor the environment is responsible for a creative and innovative response which may take place in any field of social endeavour may be business, agriculture, social work and education and the like. For the entrepreneur, it is important to have knowledge about the economic and political environment, more particularly about the economic policies of the government and the financial as well as commercial institutions. As a result, entrepreneurship can occur in a range of environments. However, the acquisition of competent characteristics for successful entrepreneurship is influenced to some extent by one's environment.

The key components of entrepreneurship is opportunity recognition, innovation, risk, use of resources, action and adding value for profit maximization where an entrepreneur is an actor who innovates by recognising opportunities, he or she makes moderately risky decisions that lead into actions regniring the efficient use of resources and contribute an added value.

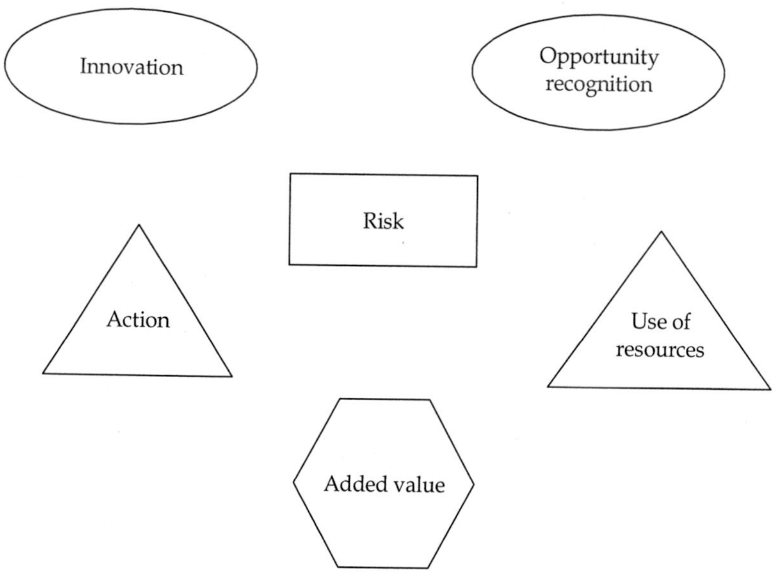

Fig. 2. Main elements used to define the 'term entrepreneur' (Filion L.J., 2008).

Entrepreneur and Manager

There are so many differences between managers and entrepreneur. Manager delegates and supervises the subordinates while having least direct involvement. The entrepreneur has more of direct involvement with least delegation i.e. entrepreneur believes in direct involvement with least delegation. Manager avoids risk while entrepreneur is ready to accept risk. The managers are mostly concerned about the likes and dislikes of their senior managers, the entrepreneur serves themselves and for their organization. Entrepreneur builds strong relations with everybody around him, whereas manager follows the relationship as per the organization chart. The prime difference between them is related to their motivation, managers are motivated by salary raise, promotion and other rewards, while independence and scope of working creatively motivates entrepreneurs.

Competencies and Characteristics of Entrepreneur

Entrepreneurship Development Institute of India, Ahmedabad identified following competencies. They are:

1. Initiation: Take action goes beyond to start a new enterprise.
2. Scanning and acting on opportunities: Utilizes the opportunities for either personal or business growth to obtain finance, equipment, or other resource.
3. Persistence: Always ready to take extreme efforts to overcome the obstacles that prevent him or her from reaching the goal.
4. Information seeking: Ready to learn from the experience of others and seek the additional information through network and other sources.
5. Work commitment: Take personal sacrifices and additional efforts to accomplish the objectives.
6. Efficiency orientation: Find base to do things faster or with fewer resources.
7. Systematic planning: Formation of logical and realistic business plan to achieve goal step-by-step.
8. Problem solving: Identification of new and potentially unique plans to reach goal and ability to switch to alternative strategies.

9. Self-confidence: Strong belief in self and own abilities.
10. Assertiveness: Direct confrontation of problem, communicate what expected of others and addressing those who fail to perform as expected.
11. Persuasion: The ability to succeed in persuading others to do what she wants.
12. Efficient marketing: Make sure that the work is accomplished according to the standards set forth.
13. Concern for employees: Keep concern and take proper action to improve the welfare of the employees.

Success of Entrepreneurship

Following aspects are necessary for the successful entrepreneurship:

1. Regular inflow of information related to buyers, consumers, distributors, dealers, retailers, transporters, etc., about raw material, quality aspects, government organisations, employees and competitors.
2. Satisfying the needs of customers.
3. Generation of adequate cashflow.
4. Regular objective assessment of the enterprise.
5. Improving productivity.
6. Maintenance of quality.
7. Use of technology of the time.
8. Be Innovative.
9. Keep employees motivated.
10. Scrap or waste material be utilised properly.
11. Time management.

Benefits of Entrepreneurship

Entrepreneurship has following three benefits for society:

1. **Economic Growth:** These provide economic upliftment of society and generate labour employment.

2. **Productivity Improvement:** It helps in improving the productivity, which means the ability to produce more goods and services with less labour and other inputs.
3. **New technologies, products and services:** It helps in promoting innovative technologies, products and services.

Training institutes involved in entrepreneurship development

1. The Entrepreneurship Development Institute of India (EDI), Ahmedabad, Gujarat - It helps in promoting micro enterprises at rural level and improvement of managerial capabilities of small scale industries. It also provides support system to facilitate potential and existing entrepreneurs establish and manage their enterprises.
2. The National Institute for Entrepreneurship and Small Business Development (NIESBUD), Noida, Uttar Pradesh. This institute is engaged in training, consultancy, research, etc. in order to promote entrepreneurship. The major activities of the institute are Training of Trainers, Management Development Programmes, Entrepreneurship-cum-Skill Development Programmes, Entrepreneurship Development Programmes and Cluster Intervention.
3. Indian Institute of Entrepreneurship (IIE), Guwahati- Designing and organising training activities for different target group and undertaking research relevant to entrepreneurship.
4. National Institute of Food Technology Entrepreneurship and Management (NIFTEM), Haryana. The main aim of this institute is "Skill Development and Entrepreneurship Development" for the food technology sector.
5. Central Institute of Tool Design, Hyderabad- MSME-Tool Room, Hyderabad (Central Institute of Tool Design) is a poineering Institution in the field of Tool Engineering in the country.
6. Kerala Institute for Entrepreneurship Development (KIED)-KIED is a premier institute promoted by Government of Kerala and Government of India for promotion and development of entrepreneurial activities in the state.
7. Kerala State Self Entrepreneur Development Mission (KSSEDM)- The main activities is to inculcate entrepreneurship development, provide training and avail finance from banks and other financial institutes.

8. Small Farmer's Agribusiness Consortium (SFAC) - To undertake or assist in undertaking programs for employment generation, growth and diversification of agriculture and agro-based industries, to identify and promote post-harvest processing units.

Financial institutes involved in EDP

State Bank of India-provide financial assistance to all categories and professionals:

1. Industrial Finance Corporation of India (IFCI)-1948- Provide schemes for agriculture,animal husbandry and entrepreneurship development scheme.
2. The Industrial Development Bank of India (IDBI)-1964- Provide schemes for the medium and small scale sector.
3. The Industrial Credit and Investment Corporation of India (ICICI)-1955- Assist in creation, expansion and modernization of the industries.
4. National Bank for Agriculture and Rural Development (NABARD) - 1982 - To provide credit for the promotion of the agriculture, small scale industries, handicrafts and other rural crafts.
5. The Small Industries Development Bank of India (SIDBI) - 1996. Financing and promotion of the small scale industries.
6. National Small Industries Corporation (NSIC)-1955-Promotion and development of small scale industries, financial assistance, training, providing raw materials for manufacturing and marketing of finished products of small scale industries.

References

Akhouri M.M.P. Entrepreneurial Economic Success, Index for Assessing Entrepreneurial Success, SEDME, 1979; IV(1): 36-39.

Cantillon, R., 1755. *Essai sur la nature du commerce en général*, London: Fetcher Gyler. Also: Edited with an English translation by Henry Higgs, London:MacMillan (1931). The manuscript was probably written around 1720 and was published after Cantillon was murdered in 1734. It is believed that he himself wrote the French and English versions.

Cole A.H., 1959. Business Enterprise in its social setting. Harvard University, Cambridge, 7.

Davidsson, P. and Wiklund, J., 2007. Levels of analysis in entrepreneurship research: Current research practice and suggestions for the future. In *Entrepreneurship*, Springer, Berlin, Heidelberg, 265p.

Say, J.B., 1815. Cathéchisme d'économie politique, Maison Mame (1972); also translation: Catechism of political economy: on familiar conversations on the manner in which wealth is produced, distributed and consumed by society, London: Sherwood (1816).

Say, J.B., 1996. Cours d'économie politique et autres essais, Paris: GF-Flammarion (Selection of main writings).

Schumpeter, J.A., 1928. 'Des Unternehmer', in Ludwig Elster *et al.* (Eds.) *Handworterbuch der Staatsvissen-schaften*, (fourth edition, Jena, 1928). In: Hartmann, H. (1959) 'Managers and entrepreneurs: a useful distinction', 430-431, *Administrative Science Quarterly*, 3(3): 429-451.

Shane, S. and Venkataraman, S., 2000. The promise of entrepreneurship as a field of research. *Academy of management review*, 25(1): 217-226.

Shapero, A. and Sokol, L., 1982. The social dimensions of entrepreneurship, Good will publications, 477p.

Stevenson, H.H. and Jarillo, J.C., 2007. A paradigm of entrepreneurship: Entrepreneurial management. In *Entrepreneurship*, Springer, Berlin, Heidelberg, 170p.

Filion, L.J., 2008. Defining the Entrepreneur Complexity and Multi-Dimensional Systems Some Reflections Rogers J.A. Bombardier Chair of Entrepreneurship, HEC Montréal.

❑❑❑

Chapter 2

The Entrepreneurial Process Approach for Developing Agricultural Entrepreneurship

Dr. P. Sethuraman Sivakumar, Dr. Jagannathan and **Dr. Sheela Immanuel**

Section of Extension and Social Sciences, ICAR – Central Tuber Crops Research Institute, Sreekariyam, Thiruvananthapuram-695 522

Indian agriculture is the second largest in the world in terms of its output and contributes 14.4% of national GDP and employs about 48.9% of the total workforce (Ministry of Finance, Government of India; Economic Survey, 2010-11). In the post-economic liberalisation era, Indian agricultural has structurally transformed, became diverse and stronger, and acquired a market-oriented business dimension. In the past few decades, India has shifted its focus on high value segments, such as dairy, horticulture and inland fisheries, which provide farmers with year-round income and account for 60 per cent of India's agricultural GDP. In the recent years, India stood as second largest producer of food and 14th largest exporter of horticultural and forestry products in the World. These developments boosted the conditions for nurturing agricultural entrepreneurships, both at the grassroot and regional levels.

Agricultural entrepreneurship is one of the fast emerging sector in India, due to the diverse business opportunities, increasing fund availability for agriculture and food sectors, easy access to quality mentoring and government support for aspiring entrepreneurs. Currently, India has over 1000 agristartups, 4000 Farmers Producer Organisations (FPO), and several thousands of small scale industries. The investment in agricultural entrepreneurship is steadily increasing, as evidenced by the steep increase in Foreign Direct Investment in the food processing sector rose from Rs. 3357 crore in 2014-15, to 4732.28 crore in 2016-17 (Press Information Bureau, July 2017). In this context, it is essential to understand the entrepreneurship development process especially in the agricultural sector.

Agricultural entrepreneurship

Entrepreneurship is the process of creating something new with value by devoting time and effort assuming the accompanying financial, psychic and social risks and uncertainties and receiving the resulting rewards of monetary and personal satisfaction. The agricultural entrepreneurship is undertaken by farmers, women, youth, groups, organisations for the purposes: (i) to stabilise market prices of agricultural commodities, (ii) to get assured income from farm produce, (iii) to generate additional income by utilising farm produce, (iv) to utilise the additional revenue or surplus money for the profitable business, (v) to generate income to sustain family and (vi) to achieve sustainable and equitable development. In general, the agricultural entrepreneurships fall under following categories:

a. Agripreneurs

The agripreneurs are primarily the farmers who are engaged in entrepreneurial activities associated with their farm. The agripreneurship development focuses on creating a new breed of farmers with core business skills in undertaking farm-based businesses for maximising their income.

According to Alsos *et al.* (2003), there are three types of agripreneurs: Pluriactive farmer, Resource-exploiting and the Portfolio Entrepreneur.

1. Pluriactive Farmer

They derive a reasonable proportion income from the off-farm income generating activists. The purpose of engaging in off-farm economic activities is to sustain their farming and/or to expand their farms to provide employment to their family members. This approach is used as a coping mechanism to sustain in adverse climatic conditions and other shocks which affect their livelihoods (Shucksmith *et al.*, 1989). In pluriactive approach, the farm business is owned by the family and less capital intensive.

2. Resource Exploiting Entrepreneur

They are the farmers who utilize the unique resources available in their farm to develop a new farm-based business. For example, the livestock farmers can prepare compost from cowdung; farm can be used as an agri-tourism venue to generate additional income. The capital requirement for the business activity varies with nature of the business.

3. The Portfolio Entrepreneur

They are progressive farmers who wish to exploit a novel but risky business idea. They develop teams to implement their ideas and ready to invest large capital for translating them into viable business. Though the ideas originate from farm, the new business is registered as a separate entity from farm. For example, when a group of farmers create a mango pulp processing factory using their own produce at the initial stages, they procure from others when expanding the production.

b. Small Business Enterprises

The small-scale enterprises are focusing only on few commodities or services. They are created by agripreneurs or rural youth to sustain their livelihoods. This business doesn't require specialised skills and can run with farmers' own capital. The examples include Agri-clinics and horticultural nurseries.

c. Entrepreneurship through Self-Help Groups (SHGs)

A SHG is a village-based financial intermediary usually composed of 10–20 local women or men. The SHGs are formed by NGOs and financed by banks to undertake a specific entrepreneurial activity. The SHGs are working mostly on traditional agri-businesses and the profit earned are utilised in a collective way.

d. Farmers Producers Organisations (FPOs)

The FPO consists of collectivization of producers, especially small and marginal farmers so as to form an effective alliance to collectively address many challenges of agriculture, such as improved access to investment, technology, inputs and markets. This collective can be registered as a company in the Company's Act and undertake farm-based business.

e. Agri-startups

Agri-startups are baby companies, which are developing innovative products or services based on a marketable idea, but yet to establish a concrete business model. They are often registered as a Private Limited Company. Any venture will be considered as a startup for the first seven years from the date of its incorporation/registration and has annual turnover maximum of Rs. 25 crores.

f. Social Enterprises

A social enterprise is an organization that applies commercial strategies to maximize improvements in financial, social and environmental well-being of people, and maximizing social impact alongside profits for external shareholders. Social enterprises are not volunteer organizations in that they operate as an enterprise by selling in a market (profit or non-profit enterprises).

Basically, an entrepreneurial process model identifies various stages and events in the entrepreneurial or new venture creation process and explains various factors and their influence on these events. It covers various stages spanning from idea generation, implementation and growth of a venture.

Stages in Entreprenaurial Process

The entrepreneurship and new venture creation research has indicated the emergences of three basic approaches of entrepreneurship (Deakins and Freel, 2006), which influence the entrepreneurial process.

- Economic approach – The entrepreneurship is viewed as a process of creating economic outputs, which focuses on the relationship between entrepreneur and economic development. Cantillon's primitive circular flow model (Brewer, 2012) is a classic example of the economic model of entrepreneurial process.
- Psychological trait approach – This approach assumes that entrepreneurship is driven by the psychological factors of the entrepreneurs than economic factors. It focuses on how an individual entrepreneur's psychological attributes such as need for achievement, internal locus of control, risk- taking ability along with others are driving the entrepreneurial process (Robertson *et al.,* 2003). The Organising model of entrepreneurship (Shook *et al.,* 2003) and Model of the Entrepreneurial Process (Brazeal

and Herbert, 1999) are few examples of psychological trait based entrepreneurial models.

- Social behavioural approach-This approach emphasizes the relationship between the external environment and personal characteristics and how they influence the entrepreneurship process. The entrepreneurial process model proposed by Bygrave and Zacharakis (2008) is a popular model representing socio-behavioural approach.

The entrepreneurial process varies with its level (e.g. macro and micro), actors (e.g., individual, team, corporation, community), by type of organization (family, small business, corporate, franchise), or by geographic location. In this chapter, we will focus only on the levels, especially micro-level process involving various actors, organisations at different geographical locations.

The macro models describe the entrepreneurial process in the regional/national level and indicate the influence of infrastructure, institutions, human resources, market forces and government policies on entrepreneurship development. Few examples include Baumol's entrepreneurial process model (Sobel, 2008) and model of endogenous economic development/stagnation (Laukkanen, 2000). The micro models focus on the individual entrepreneur's psychological attributes and other micro processes which influence venture creation. The Harmonized model of entrepreneurial Process (Hindle, 2011) and Shane's entrepreneurial process (Shane, 2003) are few models dealing with micro processes of entrepreneurship development.

Model of the Entreprenaurial Process

Bygrave and Zacharakis (2008) developed an entrepreneurial process model-based on Carol Moore's approach (Fig 1). This model indicates all the functions, activities, and actions involved in entrepreneurial process from perceiving opportunities to creating organizations to pursue them.

The entrepreneurial process model postulates that the entrepreneur passes through four events, such as innovation, triggering event, implementation and growth for creating ventures. Various personal, sociological, organisational and environmental factors influence each event.

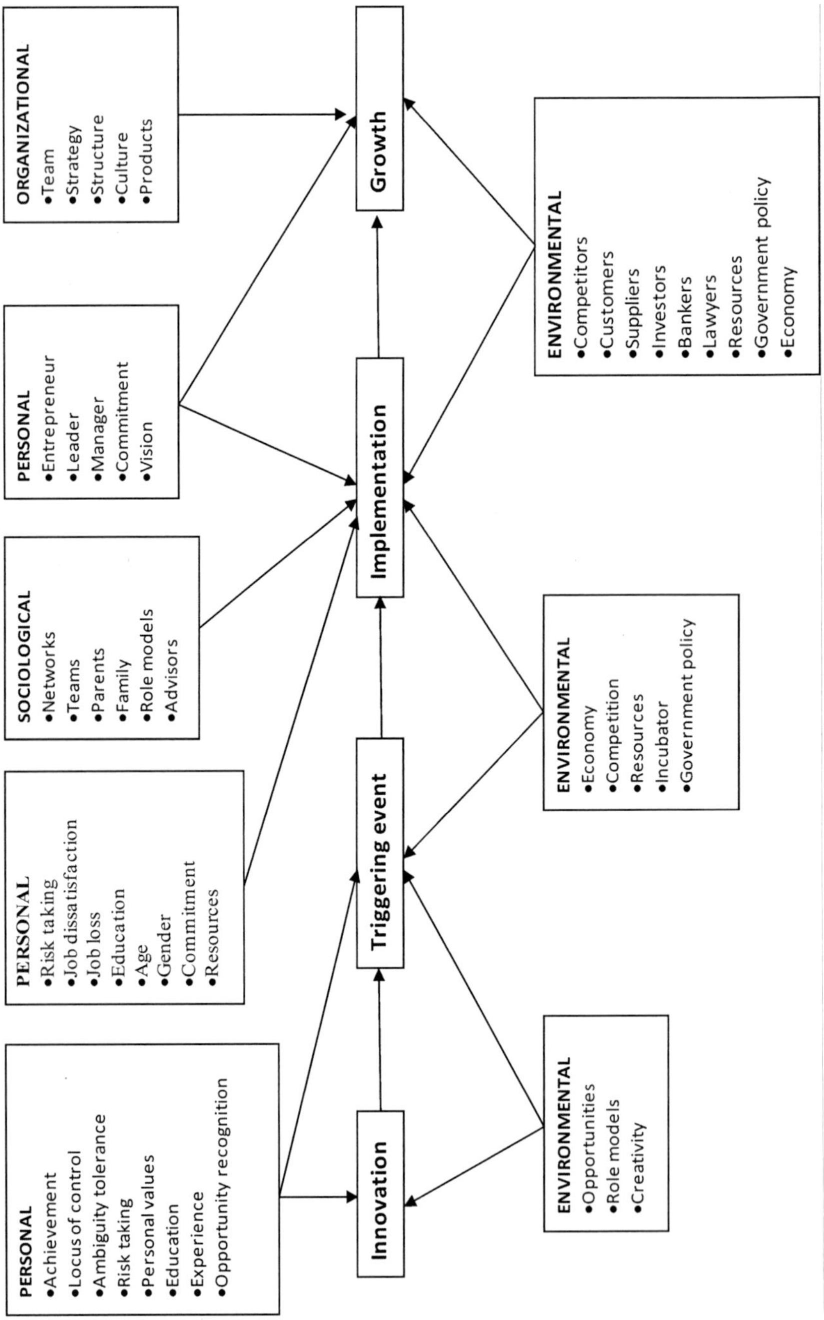

Fig.1. Entrepreneurial process [Adapted from Bygrave, and Zacharakis (2008)]

1. Innovation

In the innovation stage, the entrepreneur develops an innovative idea, locates the market opportunity, and begins to seek information to assess the opportunity. He/ she also test the feasibility of the business idea and opportunity, and his/her ability to get value from it, and how to generate the development of the product or service.

The business idea and opportunity are the key aspects to covering an innovation into a marketable product/service. The differences between idea, business idea and business opportunity are displayed in Table 1.

Table 1: Distinguishing characteristics of idea, business idea and business opportunity

Idea	Business idea	Business opportunity
1. A thought or collection of thoughts generated in the mind	1. An idea which is used to make money	1. Proven business idea that generate income
2. Generated intentionally or unintentionally	2. Intentional	2. Intentional
3. Economic aspects are not attached	3. Centred around a product or service which makes money	3. Proven money making aspect

If the entrepreneur is planning for technology-based entrepreneurship, his/her progress through this innovation event is influenced by the stage of agricultural technology.

Stage 1 - Early stage : An early-stage technology is an idea which is expected to work and solve an existing problem or create a new need. For example, a plant extract known to have a pesticidal property, but the components and modalities are unknown.

Stage 2 - Proof of concept : The early stage idea was researched and a new technology has been developed to the point that it shows signs of having the proposed effect. In this stage, few components of the plant extract which cause the insect morality have been identified, but the mechanism by which they act is unknown.

Stage 3 - Reduction to practice: In this stage, several experiments on the specific idea have been completed and the intended results have been reliably and repeatedly reproduced. The pesticidal properties of specific components of the plant extract have been identified and mode of action is documented and validated.

Stage 4 - Prototyping, formulation and compound: The technology is now standardised and found reliable and valid and ready for commercialisation. In the previous example, the components having pesticidal properties are extracted using a specific method and reformulated into a pesticide with target-specific claims. Source: Markman *et al.* (2005).

2. Triggering event

Triggers are forces in the individuals or in the individuals' perception of their situation that move them towards the entrepreneurial process (Liang and Dunn, 2007). The triggering events are incidences happened as an occasion or an episode that influence individual's assessment on the situation. Such events happen during the gestation process, during which the entrepreneur develops motivation to start a business and to decide to proceed with. At this stage, he/she develops feasibility report and business plan, as well as the identification of the resources required, estimation of the project risk, determining the source of the funds and how they would use them.

The feasibility report indicates the viability of the business idea - market viability, technical viability, business model viability, management model viability, economic and financial model viability, and exit strategy viability. The business plan converts the feasibility report of a business opportunity into a workable form by elaborating on the tactics and strategies to be implemented to start and grow the business.

Busiess Plan

A Business Plan is a written statement that describes and analyse the proposed businesses and provides detailed projections for the future. A business plan is bankable only when it has sufficient collateral, projects future cashflow, and indicates high probability of success and is acceptable to institutional lenders or banks or investors for providing financing.

In general, business plans are created for: (i) securing funding for the proposed business, (ii) Helps the proposer to decide to proceed or stop, (iii) Helps to improve the business idea , (iii) Improves probability of success and (iv) Helps to keep the business on track. Various elements of business plans are given.

The personal attributes like achievement motivation, locus of control, ambiguity tolerance, risk-managing, personal values, education experience, Job dissatisfaction, Job loss and ability to recognise opportunities influence the innovation and triggering events. The sociological factors networks, teams, parents, family, role models and advisors as well as environmental factors like opportunities, role models and creativity also influence the innovation and triggering events.

3. Implementation

The implementation event includes registration and clearances, pooling of resources, man power and other infrastructure to launch the new business to the market. The business strategy, day-to-day operation, and other operations develop gradually for building and managing the venture. Securing funding and establishing the venture are three major components in the implementation process.

1. The Team

a) Who are you? - Team

b) Your credentials? – Qualifications and experience

c) Faculty or other personnel who helped you to develop this plan? - Mentor

2. Idea / Opportunity

a) What's your idea? – Describe the business idea

b) Provide a technology/service overview-Describe the technology/service

3. Problem / Pain/Need

a) What problem is it solving? – Description of the problem it is addressing

b) How significant is the problem for the country? - Extent

4. Market

- What is the market for your idea? – Describe the market, market size.

- How would you categorize competing technologies in the market and where does your idea fit? – Unique value proposition.

5. Customers

- Who are your target customers (i.e. target market)? - Draw and describe (Segmentation)
- How will you get your product to them? - Marketing channel

6. Competition Assessment

a) Who are your competitors in this category?

b) What are their strengths & weaknesses?

c) Are the competitors' sales growing or declining?

d) What are your competitive advantages? (That is, why should you be selected by customers instead of the competition?)

7. Product Development

- What is your product development plan (from concept to prototype to production to market)? What is the timeframe?
- Are there key go/no-go decision points?

8. Intellectual Property

a) How will you protect your invention & when will you initiate the intellectual property protection in your product plan?

9. Commercialization alternatives

a) Licensing to a company in this category?

b) Starting a company? Risks vs. Rewards.

10. Financing

a) If starting a company, what are projected costs and revenues? How might you finance it initially?

b) Costs of IP protection?

11. Conclusion

a) Is this a true opportunity or not?

b) What did you learn in conducting this analysis?

(i) Securing funds

Securing funds for the new ventures are difficult part of the entrepreneurial process. At the initial stages, the entrepreneurs need seed funds to convert their idea into a marketable product and initiate the venture formation process. The entrepreneurs often invest their own, family resources and funds from friends/relatives to develop the venture. This mode of financing is called as boot strapping, which is a common form of financing ventures in India. In general, there are two major sources of financing new ventures like debt and equity financing:

A. Equity Financing

The equity funding is an umbrella term that refers to any means of financing a company in which the owners receive money in exchange for sharing a part of ownership. There are multiple methods for raising equity capital ranging from 1 to 100%, depending upon the nature of business. Types of equity financing include self-financing/bootstrapping, Friends and family members, government grants, angel Investors, crowdfunding and venture investors.

B. Debt Financing

The debt financing refers to borrowing funds from creditors with the stipulation of repaying the borrowed funds plus interest at a specified future time. The types of debt financing include friends and relatives, banks and other commercial lenders, commercial finance companies, government programs, bonds, etc.

Currently, many entrepreneurship promotion agencies, incubators, angel investors' offer seed funding to the aspiring entrepreneurs. Several grants and incentives are also available from public and private sector agencies for new venture formation.

(ii) Establishing venture

The establishment of venture starts after securing initial funding. It involves identifying location, construction of buildings, procuring equipments and creating facilities, hiring quality human resources, sourcing raw materials, establishing partnerships for various functions, implementing marketing plan and finally making the products or services available to the customers.

The sociological factors networks, teams, parents, family, role models and advisors as well as environmental factors like economy, competition, resources, incubator and Government policy influence the implementation stage of the entrepreneurial process.

4. Growth

Growth event of the entrepreneurial process occurs, when an entrepreneur operates his/her established business and begin to get payback for efforts. In this phase, the entrepreneur runs daily operations, manage issues, getting customer feedback of products/ services and monitoring sales and services. Based on the success or failure of his/her efforts, the entrepreneur decides on the future prospects of the business. The actual growth is compared against the planned growth and then the decision, regarding the stability or the expansion of business operations is undertaken by an entrepreneur. In general, the growth phase is critical for agricultural ventures as over 70% of them fail in the first three years of inception due to inadequate funding, poor management, and market failures. Thus, it is essential to develop sound business ideas, backed up with strong business plans with adequate foresight.

The organisational factors like the expert team, strategies, and structure of the venture, culture within venture and the products/ services influence the growth phase of the entrepreneurial process. Several environmental factors like competitors, customers, suppliers, investors, bankers, lawyers, resources, Government policy and national economy influence the growth stage.

Conclusion

The agricultural entrepreneurship is one of the promising areas in agriculture, and its growing exponentially in the recent years. With increasing value generation and investment, they provide profitable alternatives for farmers and rural youth who were deprived of by middlemen for years. Though the entrepreneurial ecosystem provides a favourable climate for generation of agricultural ventures, their growth is halted by widespread failure. Thus, it is essential to understand the entrepreneurial process adequately and act upon the critical aspects to maximise the success of the agricultural ventures.

References

Alsos, G.A., Ljunggren, E., Pettersen, T.L., 2003. Farm-based entrepreneurs: what triggers the startup of new business activities? *Journal of Small Business and Enterprise Development,* **10:** 435-443.

Brazeal, D. and Herbert, T., 1999. The genesis of entrepreneurship. *Entrepreneurship Theory & Practice,* **23(3):** 29-45.

Brewer, A., 2012. An Essay on Economic Theory: An English Translation of Richard Cantillon's Essai sur la nature du commerce en général. *History of Political Economy* **44(4):** 693–695.

Bygrave, W.D., and Zacharakis, A., 2008. *Entrepreneurship*. Hoboken, NJ: John Wiley & Sons.

Deakins, D. and M. Freel, 2006. *Entrepreneurship and Small Firms*, Berkshire, UK: McGraw-Hill.

Hindle, K., 2010. Skillful Dreaming: Testing a General Model of Entrepreneurial Process with a Specific Narrative of Venture Creation. In *Entrepreneurial Narrative Theory Ethnomethodology and Reflexivity*, ed. W. Gartner, 97-135. Clemson, SC: Clemson University Digital Press.

Laukkanen, M., 2000. Exploring alternative approaches to high-level entrepreneurship Education: creating micro mechanisms for endogenous regional Growth. *Entrepreneurship & Regional Development,* **12(1):** 25-47.

Liang, C. and Dunn, P., 2007. Triggers of the decision to launch a new venture - Is there any difference between pre-business and in-business entrepreneurs? *Academy of Entrepreneurship Journal,* **13(1):** 79-88.

Markman, G.D., Phan, P.H., Balkin, D.B., and Gianiodis, P.T., 2005. Entrepreneurship and university-based technology transfer. *Journal of Business Venturing,* **20:** 241–263.

Ministry of Finance, 2011. *Economic Survey, 2010-11*. Govt of India.

Press Information Bureau (July 2017). *FDI in Food Processing Industries.* Retrieved from: http://pib.nic.in/newsite/Print Release.aspx ?relid=168879.

Robertson, M., A. Collins, N. Medeira and J. Slater, 2003. Barriers to start- up and their effect on aspirant entrepreneurs, *Education & Training,* **45**(6): 308–16.

Shane, S., 2003. *A General Theory of Entrepreneurship. The Individual opportunity nexus*. Cheltenham (UK): Edward Elgar.

Shook, C.L., Priem, R.L. and McGee, J.E., 2003. Venture creation and the enterprising individual: a review and synthesis, *Journal of Management*, 23(3): 379-399.

Shucksmith D.M., Bryden J., Rosenthall P., Short C., Winter D.M., 1989. Pluriactivity, farm structures and rural change. *Journal of Agricultural Economics*, **40**: 345-360.

Sobel R.S., 2008. Testing baumol: Institutional quality and the productivity of entrepreneurship". *Journal of Business Venturing* **23**: 641-655.

❑❑❑

Chapter 3

Finishing School - A Skill Sharpening Methodology for Enhancing the Employability and Entrepreneurial Skills

Dr. A. Anil Kumar (Faculty Dean), **Karishma Zen** and **Namitha Raghunath** (Research Scholars)

Department of Agricultural Extension, Kerala Agricultural University, College of Agriculture Vellayani, Thiruvananthapuram-695 522, Kerala

Introduction

Finishing school programmes are mainly held by different institutions for familiarizing the students with professional networking skills, which they can use to find jobs for themselves and others. Thus, many countries look to improve their human resource development strategy to take advantage of certain growing industry that seems to have room for many players, aiming at bridging the gap between formal education and industry requirement. Roles (2015) claim that these finishing school programme not only enriches the skill of students, but also improves their entrepreneurship abilities and instills confidence to venture into successful business after their school tenure. Finishing school was evolved as "charm school" a school for young women that focuses on teaching social graces and upper-class-cultural rites as a preparation for entry into a society (Garcia and Bafundo, 2014).

Finishing school programmes are strategic training and development tool to find a shortcut for the employability of human resources within the global services industry in developing countries (Gereffi *et al.*, 2011). Countries that want to develop and increase their share in the global outsourcing market must be cognizant of the necessity of counting with a scalable and employable labour pool capable of performing the outsourced tasks.

The main idea of a finishing school programme is to enhance capacity, employability and public-private partnerships of the labour pool through a complementary education and training framework that aims at supplementing, rather than substituting, formal education. The concept of "finishing" implies the existence of a necessary previous step involving formal education; however, as it is naturally difficult to keep up with the dynamic and changing market forces that drive global services, finishing schools are designed to be a more flexible complementary tool that can tackle those dynamic forces and train employees according to the latest industry requirements, programme consist of practical knowledge regarding the specific field of work, a committed attitude towards continuous learning, soft skills, team work skills and analytical capabilities.

Prospects of Vocational Education

Vocational education, or essentially be termed as a job-oriented education round the globe, intends to empower the clients by enhancing their competencies so as to enable them to become better suited in job markets, including the ones newly emerged in the wake of the faster pace in growth and development of the society. In an agrarian society, vocational education in agriculture assumes greater relevance. The Government of India also realized the importance of vocationalisation and attempted to launch vocational education at the higher secondary level. Despite all these efforts, it is felt that the intervention of vocationalisation of education has not yielded intended results at a tangible rate. The main reasons responsible were lack of effective management system, absence of employment opportunity in the specific sectors in which the vocational courses were offered, mismatch between demand and supply, reluctance in accepting the concept of vocationalization at an early stage of the client's educational phase by the society, etc. In short, the expectations of the agripreneur remain largely unaddressed, mainly due to the lack of skill component of the VHSE system. The prejudice against vocational education will not, thus, be removed unless a reasonable change is occurred and sizeable employment opportunities are created to the students of vocation education by converging the different supply factors and interests of the stakeholders.

It had aided in the individual skill development in vast miscellany of fields of interest at a comparatively younger age. The new education policy clearly spells out that vocationalization should be introduced in schools and colleges all over the country at the higher secondary stage. Vocation-based or job-oriented course have been introduced not only at the higher secondary stage but also at the degree level. Several colleges in the country are running vocational courses and turning technically trained people in different fields.

Finishing School Approach in Kerala

Considering all these, the finishing school approach has been designed for sharpening and refining personal competencies of the students and to provide specialization in selected agricultural innovations for taking up agricultural enterprises successfully. Finishing schools help in soft skill training for those specific students who pass out from the last year of professional courses like agriculture, engineering, law studies, and business and management courses but lack in certain key skill areas. Students who join these schools, will get a unique opportunity to obtain the correct balance between both, theoretical studies and practical job. Advantages of finishing schools are that it encourages confidence in trainees and helps in grooming the individuals, teaches students key business etiquette and helps in interaction with senior industry level experts.

In Kerala, Resource Enhancement Academy for Career Heights [REACH] is a finishing school run by Kerala State Women's Development Corporation, which support the woman population with professional training, guiding them towards a bright future. In addition, the Model Finishing School was the combined effort of the IT Department, Institute of Human Resource Development and Government of Kerala to help the youth to grab the best career by providing them with skill development in various aspects.

Kerala, has been underscored the most literate state in the country. This routes way back to the antiquity of education in the state. Education system in Kerala has been through so many hurdles from the past british rule to the freedom era only to strengthen itself out. A plenty of well-developed schools and colleges lead Kerala to today's phase. Kerala became the first Indian state to achieve 100 per cent primary education through its literacy programme *Athulyam*. Likewise so many e-projects has been popularized in the state for the first time in the country. But in spite of the high literacy, unemployment is the major crisis faced by the people of Kerala. The major problem in Kerala's education sector is the lack of skill development. The

Government realized the significance of vocationalisation and took great effort to launch vocational education at the senior secondary level. Vocational education was intended to educate and develop expertise in technology and scientific skills in order to give their full potential in their area of interest. Vocational education has been put to form on a number of bases, such as commerce based, engineering based, health and para medical based, home science based, agriculture based and hospitality and tourism based. These streams provide the necessary essentials in a skill development platform to excel in these.

Finishing School for VHSE (Agri.) Certificate Holders by Kerala Agricultural University

Kerala agricultural University (KAU), had the farthest view and thought to design a finishing school methodology and was brought to action in the year 2011.The major objective of the programme was to sharpen and refine the personal competencies of students, to provide specialization in selected agricultural innovations and to transform the trainees into model frontline agripreneurs.

Specific Objectives

- Need assessment of VHSE plus two certificate holders in viable vocations in agricultural sector based on the field requirements and market demand.
- Identification of viable agricultural enterprises/vocations utilizing hi-tech Agri innovations suitable to the socio-political agricultural context of Kerala.
- Exposure to the selected cafeteria of hi-tech agricultural enterprises/innovations for selecting a set of enterprises according to the need of the candidate and relevance of the context, and providing skill-dominant training in selected agricultural innovations through Finishing School programme.
- Providing apprenticeship support to the candidates who have completed finishing school programme and willing to take-up selected vocations as enterprises.
- To develop network of candidates to address the day-to-day requirement of vocations.
- To document the key activities and performance of the project.

Monitoring and evaluation of the components at different stages of the project

So far seventeen batches had successfully completed their programme from three main centres, namely College of Agriculture, Vellayani, College of Agriculture, Padannakad and Centre for e-learning, Vellanikkara. The programme included sixty days training programme, followed by an apprentice programme of six months. In the first two months of training focuses on sculpting the inbuilt and acquired skills, and also to impart more skills to the students. The so called skill training phase is mainly focused on five modules *viz.* soft skill, entrepreneurship development, enterprise attachment, computer skills and project formulation and planning. The six months apprentice programme focus on allotment of candidates to different departments in the university and state department of Agriculture, for which they are awarded a stipend of Rs. 6000 per month. The participants are selected by a team of scientists led by the project leader after conducting a preliminary written test. The final selection is based on interview.

The details of modules are presented below:

S.No.	Module	Duration	Purpose
1	Soft skill	15 days	Give exposure to the VHSE certificate holders to improve their efficiency and elevate their self confidence through enhanced communication skill, problem solving skills as well as gain ability in risk management and manage time as a limited resource.
2	Computer skill	10 days	Develop ICT enabled communication cum information processing skill among the participants.
3	Entrepreneurship development	15 days	Capacity building of the participants as an entrepreneur in the various dimensions of agri enterprise encompassing planning, project formulation, pilot testing, promotion and marketing, as a best team player to achieve common and self goals towards success.
4	Enterprise attachment and feedback	10 days	Provide hands on exposure to the participants in various identified agri-enterprises and selecting a group based viable project from among the enterprise in which he had got exposure.
5	Project formulation and report preparation	10 days	Provide hands on exposure to the participants in various identified agri-enterprises and gaining experience in the management of the enterprise and selecting a group based viable project from among the enterprise in which he got exposure.

(Finishing School report, 2015)

Other Initiatives by Government

1. ARYA (Attracting and Retaining Rural Youth in Agriculture)

In order to create interest and confidence among rural youth in agriculture, ICAR has initiated a new scheme ARYA (Attracting and Retaining Rural Youth in Agriculture). This program is planned to be implemented through KVKs in 25 states of the country. Each KVK trains about 200 to 300 youth in taking up agriculture and allied activities like poultry farming, dairying, fisheries, mushroom production, etc. which keep the rural youth attached to agriculture, either directly or indirectly. Finally, the trained young entrepreneurs would be assisted in preparing project reports for seeking bank loans.

The objectives of ARYA project are :

- To attract and empower the youth in rural areas to take up various Agriculture, allied and service sector enterprises for sustainable income and gainful employment in selected districts.
- To enable the farm youth to establish network groups to take up resource and capital intensive activities like processing, value addition and marketing.
- To demonstrate functional linkage with different institutions and stakeholders for convergence of opportunities available under various schemes/program for sustainable development of youth.

2. Student READY (Rural Entrepreneurship Awareness Development Yojana)

The Student READY (Rural Entrepreneurship Awareness Development Yojana) programme aims to provide rural entrepreneurship awareness, practical experience in real-life situation in rural agriculture and creating awareness to undergraduate students about practical agriculture and allied sciences. This programme will help in confidence building, skill acquisition and documentation of Indigenous Technical Knowledge (ITK) of the locality and thereby, enabling the graduates for self-employment. It also provides opportunities to acquire hands-on-experience and entrepreneurial skills. It was felt necessary to introduce this program in all the SAU's as an essential pre-requisite for the award of degree.

The following components are proposed for carrying out one year Student READY programme in all the Under graduate (UG) disciplines:

- Experiential Learning on Business Model/Hands on Training
- Experiential Learning on Skill Development
- Rural Awareness Works Experience (RAWE)
- Internship/In-Plant Training/Industrial attachment
- Students Projects

All the components as per suitability of course i.e. Experiential Learning, Skill Development Training, Rural Agricultural Work Experience (RAWE), Internship/In-plant training and Student Projects are included in the final year of study for 2 semesters to provide entrepreneurial skills, confidence and hands on experience. There are 20 credits for Experiential Learning/ Skill Development Training, 10 credits for RAWE and 10 credits for Industry Attachment/Student Project.

3. ASAP (Additional Skill Acquisition Programme)

Additional Skill Acquisition Programme (ASAP) is a joint initiative of General and Higher Education Departments, launched by Government of Kerala with an objective to impart quality skill education to the higher secondary/undergraduate students along with their regular curriculum. Owing to the success of the program, attained over the years, the target group is widened to professional colleges and is now set to kick off community skilling through the Community Skill Parks.

4. Pradhan Mantri MUDRA Yojana (PMMY)

Pradhan Mantri Mudra Yojana (PMMY) is a flagship scheme of Government of India launched on 8th April, 2015 by the Hon'ble Prime Minister to "fund the unfunded". It enables a small scale entrepreneur to avail credit upto Rs. 10 lakhs from all Public Sector, Private Sector, Foreign Banks, Micro Finance Institutions (MFI) and Non-Banking Finance Companies (NBFC) for non-farm income generating activities.

Under MUDRA, the schemes include :

- Shishu : covering loans upto 50,000/-
- Kishor : covering loans above 50,000/- and upto 5 lakhs
- Tarun : covering loans above 5 lakhs and upto 10 lakhs.

The interventions have been named 'Shishu', 'Kishor' and 'Tarun' to signify the stage of growth/development and funding needs. It would be ensured that at least 60% of the credit flows to Shishu Category Units and the balance to Kishore and Tarun Categories.

5. Pradhan Mantri Kaushal Vikas Yojana (PMKVY)

Pradhan Mantri Kaushal Vikas Yojana (PMKVY) is the flagship scheme of the Ministry of Skill Development & Entrepreneurship (MSDE). The objective of this Scheme is to enable a large number of Indian youth to take up industry-relevant skill training that helps in securing a better livelihood. Under this Scheme, Training and Assessment fees are completely paid by the government.

Conclusion

Finishing school programmes have paved the way of new opportunities for our youth. In a scenario of unemployment, finishing school programmes combined with the initiatives of central and state governments will prove to be a boon for the unemployed educated youth of our nation. New interventions and programmes aiming for socio-economic development should be encouraged and promoted for a sustained future.

References

Bassi, M., Matías, B., Sergio, U., and Jaime, V., 2012. *Disconnected: Skills, Education and Employment in Latin America.* IDB, Washington, 128p.

Garcia, M.P. and Bafundo, F. 2014., *Best Practices in Finishing School Programs for the Global Services Industry.* Inter-American Development Bank, Washington, 41p.

Gereffi, G., Stark, F.K., and Psilos, P., 2011. *Skills for Upgrading: Workforce Development and Global Value Chains in Developing Countries.* Duke Center on Globalization, Governance & Competitiveness, RTI International, North Carolina, pp. 1-12.

□□□

Chapter 4

Agritech Startups: New Age Entrepreneurship Aiding Accelerated Growth in Agriculture

Dr. Saji Gopinath (Chief Executive Officer)

Kerala Startup Mission, Technopark, Thejaswini, G3B, Technopark, Kariyavattom

India is fast emerging as a global hub for Startups. The vast talent of science and technology professionals aided by proactive policies of Government has resulted in development of many innovative companies, which have created substantial wealth and employment while solving some of the erstwhile perennial problems of the society. These new age entrepreneurs mostly leverage on their intellectual capital to develop affordable solutions to enhance growth and productivity in various sectors of the economy. Even though, the e-commerce startups like flipkart, myntra and aggregators like Ola, Oyo, etc. are more popular, there are a large number of deep tech startups who are making significant contributions to improve the productivity of core sectors of economy like Agriculture. This chapter looks at the need, relevance of Agri-tech startups and looks at ways of developing such startups. Some of the inititives of State and Central Government to support such startups are also discussed here.

What is a Startup?

There are many definitions for startups. On a broader scale, any entrepreneurship activity that creates value can be considered as a startup. However, the Government of India has stipulated following conditions to consider an entity as startup[1] for availing benefits and concessions under the Startup India Scheme of the Government.

---

1 Gazette G.S.R. 364 (E) dated 11/04/2018, Ministry of Commerce and Industry, Government of India

Conditions for considering an entity as a Startup

i) Age of the entity to be less than or equal to seven years from the date of incorporation/registration, if it is incorporated as a private limited company (as defined in the Companies Act, 2013) or registered as a partnership firm (registered under section 59 of the Partnership Act, 1932) or a limited liability partnership (under the Limited Liability Partnership Act, 2008) in India. In the case of Startups in the biotechnology sector, the period shall be upto ten years from the date of its incorporation/registration.

ii) Turnover of the entity for any of the financial years since incorporation/registration has not exceeded Rs. 25 crore.

iii) Entity is working towards innovation, development or improvement of products or processes or services, or if it is a scalable business model with a high potential of employment generation or wealth creation.

Thus, Startup companies can be considered as entities who are working towards innovation, development or improvement of products, processes or services, or entities having a scalable business model with a high potential of employment generation or wealth creation. The innovation potential and/or capability for creating huge wealth or employment make such companies new drivers of growth across the world. It is important to understand the difference between traditional entrepreneurship and startups. While Startups is a new way of entrepreneurship, the key difference lies in the deployment of intellectual capital rather than physical capital in developing enterprises. This is well-captured in the economic theory that suggests the changing the nature of entrepreneurship in a country, as shown in Figure 1 below :

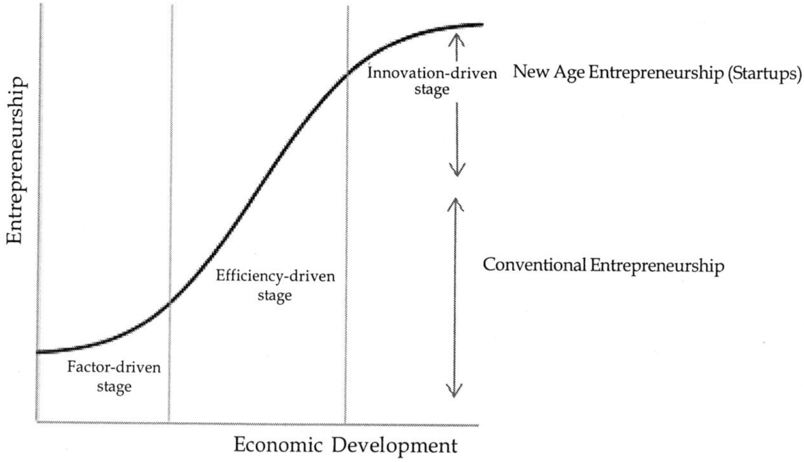

Fig. 1 : The changing nature of entrepreneurship with economic growth of a region

As shown in figure, when an economy (or a sector of economy) is growing slow, the key entrepreneurship focus will be factor-based where primary factors like land, labour, capital play a significant role. Many of the traditional business entities and agricultural farms still follow this model. As economic progress gains pace, the focus of entrepreneurship shifts to ways of enhancing the efficiency. The adoptions of farm mechanization, mechanized manufacturing, etc. are examples of this. However, in order to gain substantial economic growth, the entrepreneurship should change focus and leverage on innovations to enhance productivity. Such innovative entities (startups) have potential to bring exponential changes to societal problems through innovative deployment of technology. As Agriculture sector demand substantial improvement in growth, the relevance of innovation driven startups in this sector is quite significant.

Startups in Agriculture

As a sector, Agriculture requires substantial improvement in growth to support a large chunk of population, whose livelihood depends on it. There is a strong need to develop innovative solutions that solves many of the problems at various elements of agriculture value chain. This involves development of products and services that enhances productivity in harvesting, storage, post processing, value addition, supply chain, distribution and allied activities. Agriculture

Startups (commonly known as Agritech startups) aim to solve these issues through deployment of new generation technologies like artificial intelligence, blockchain, geospatial data analytics, etc. The common type of Agritech startups are depicted in the Figure 2 below

As shown in figure, various Agritech startups focuses on different elements of the value chain. Some of the examples of Agritech startups are discussed below :

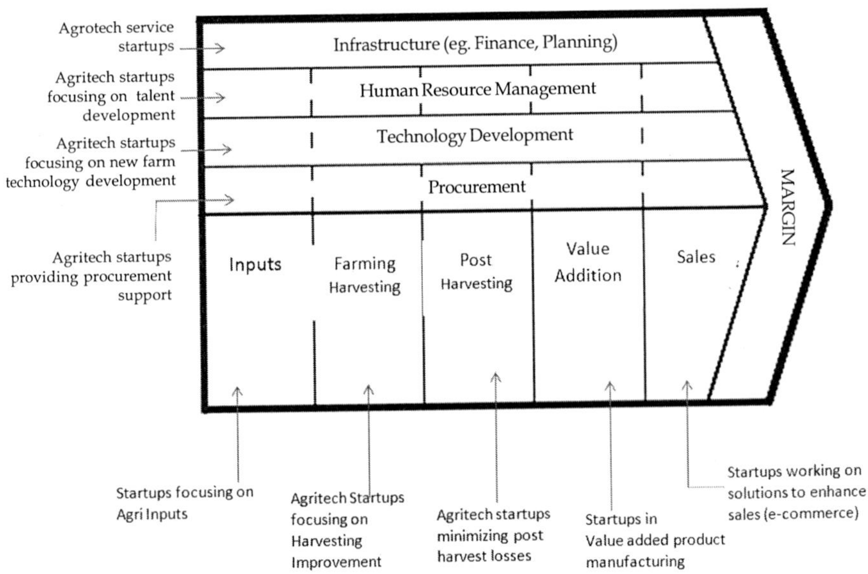

Fig. 2: Focus of Agritechstartups on various elements of Agriculture Value Chain

While there are startups in the entire value chain of Agriculture (as shown above), majority of them focus on enhancing the productivity of supply chains through technology. There are AI (artificial intelligence), based startups that provide effective detection of diseases and leverage extensively on geospatial technologies to optimize harvesting patterns to gain better price. Several startups uses traceability solutions like IoT (Internet of Things) and Blockchains to ensure farm to kitchen transparency thereby generating higher value for farmer yields. Biotech startups that works agri-inputs enhances farm productivity as a whole. There are also startups who work on e-commerce solutions to expand the market for farm produce by trying to remove expensive middle men between the producer and consumer. All these startups try to unlock the potential of farm sector through deployment of new age technologies. Some of the technologies used are discussed in the next section.

Startups and Technology

The world is in the cusp of another technology revolution. Akin to the Industrial revolution that enhanced the physical productivity of labour multifold through inducement of mechanical engineering technologies to replace human labour, the cognitive power of humans are being magnified many times through computing machines with artificial intelligence capabilities. The small machines that could process billions of data points within seconds and could tolerate ambiguity and impreciseness, would change the way people live, work and entertain in the near future. The autonomous cars, general purpose robots, fast mobility solutions, revolutionary Biocomputing devices, etc. are all are shaping the future in ways which are hitherto unknown to humankind. These technologies which converge physical, digital and biological world, commonly known as Industry 4.0, has a key role in bringing exponential growth in developing economies like India. Traditional sectors like Agriculture has lot to gain from inducement of such technologies as many perils of current agriculture like water management, pest management, nutrient management, etc. can be well-addressed by these technologies. It is pertinent to note the benefits the farm automation of third industry revolution brought to Agriculture during the 20th century. Similar to that the evolution of artificial intelligence can bring about substantive changes in the entire agriculture value chain resulting in multifold enhancement of productivity and return. For instance, the Artificial intelligent solutions coupled with video analytic solutions can effectively detect incidence of pest and diseases much better than human beings. Similarly, as discussed earlier, by Blockchain enabling of the entire farm to market chain, visibility of crops can be improved substantially. As Industry 4.0 combine digital, physical and biological worlds, comprehensive solutions can be developed to address the problems that used to plague the sector for a long time. As these technologies are still evolving, they can be better harnessed by small startup enterprises which are quite innovative, agile and responsive. This is the role of agri-tech startups. As the most literate State in the country with a large population of Science & Technology professionals, this provides a new path of development for Kerala Agriculture. The enabling environment available in Kerala for supporting accelerated growth of such startups is discussed in the next section.

Developing Agritech Startups – Enabling Environment

As discussed earlier, startups leverage substantially on their intellectual capital to develop innovative solutions of tremendous value. They are generally asset light and leverage extensively on R&D efforts. This require good amount of resources for their initial development. Unfortunately the traditional sources of finance like banks and financial institutions perceive them to be too risky and offer little support for their development and growth. For instance, of the 4.2Bn USD investment which came to Startup ecosystem in India, the share of conventional organised lending institutions and banks are only a miniscule. Hence, it is necessary to devise novel models of financing to acquire resources required for their growth and development. The new forms of alternate sources of funds like low cost seed capital, angel and VC funding is necessary for development of a startup ecosystem. Easy access to new age technologies and abundance of qualified and passionate human resources are also necessary for development of startups. There is also a need for Universities and Research agencies working in basic and applied sciences to realize the huge potential startups offer for economic growth and re-orient manpower to develop innovation and entrepreneurship.

Since most of the startups are small but work on cutting edge technologies, creating innovations which could disrupt the traditional ways of doing businesses, they require small geographical footprints making them ideally suitable for Kerala. Realising the relevance of startups as key growth engine and the primary employment provider of future, Government of Kerala was in the forefront to develop the startup ecosystem in the State. As the first state in India having a Startup policy, the Government has evolved a comprehensive framework for developing the ecosystem conducive for nurturing technology startups from Kerala. It is heartening to note that with these efforts, Kerala now have more than 1500 startups, some of them working in addressing the problems of Agriculture sector. However, given the need to enhance the productivity of Agriculture sector, there is a need to develop many more startups in this sector.

For the development of Startups in Kerala, Government of Kerala has created an apex body Kerala Startup Mission[2] (KSUM) under Department of Electronics & IT in 2014. In addition to KSUM, Kerala

---

[2] For more details visit startupmisison.kerala.gov.in

State Industry Development Corporation (KSIDC) and various Universities and Institutions like IIMK, NIT Calicut, KUFOS, KAU, KVASU, etc. are also active in developing startup ecosystem in the State. Under KSUM, a comprehensive framework for developing technology-based Startup ecosystem is being rolled out in the State.

Kerala follows a "ground up" model of startup ecosystem development. This involves implementing schemes that enable innovation and entrepreneurship right from schools. As Industry 4.0 technologies provide potential for creating solutions that opens up Blue oceans of opportunity, it is necessary that the startup entrepreneurs are provided with adequate exposure and training in these deep technologies. Towards this, KSUM has established collaboration with MIT- Centre for Bits and Atoms and set up a chain of Fab labs that provide access to modern machining techniques to people across the State to develop a much needed maker culture. The availability of new forms of manufacturing like 3D printing, laser cutting, embedded systems development, etc. have supported creation of several unique and useful products from the Startup Ecosystem. India's first and only solar-based passenger ferry which eliminated pollution and use of fossil fuels; India's first small under water remote operated vehicle (drone) used by DRDO for under water sonic experiments; the remote control drain cleaning robot that saved thousands of people from manual scavenging, etc. are all products the startups have developed with the support of this infrastructure. The creation of maker culture also helped KSUM in developing Maker Village, its electronic incubator, as the largest electronic startup workspace in the country housing over 60 companies involved in development of many novel products. It is also heartening to note that maker Village could file 24 patents during last 1 year, indicating the Intellectual competitiveness of the Startup ecosystem in the State. Many companies in Maker Village work in Agri problems. For instance, Nava Design which developed an IoT-based Neera tapping machine than enhances the tapper productivity by over 200 times, Resonva technologies that developed a milking machine that monitor animal health parameters, Fetherdyne that provides automatic monitoring of plantations are some of the examples that are creating interesting solutions to problems of agriculture.

In addition to Fab labs and Maker Village, KSUM also created a unique bio-incubator in association with Rajiv Gandhi Centre for Bio Technology to support development of hi-tech bio-startups from

Kerala. The incubator, with modern biotech tools and equipment, hosts around 20 biotech companies involved in advanced research and development. In tune with its vision of nurturing startups working in new age technologies, KSUM also has accelerators in Mobile Application Space and is also developing niche ecosystems for Space tech, Cancer Tech and Agri-tech in different parts of the State. Some of the solutions from these incubators especially from Space Tech has applications in Agri sector too.

Another intervention planned by KSUM for creating startup companies capable of excelling in Industry 4.0 is the development of Centres of Excellence in exponential technologies. CoE in AR/VR, Design, Cyber Security, Digital Fabrication, Blockchain, Cognitive Computing, Space Technologies, etc. are at various stages of implementation. These CoEs gives high end training for skills development and provide access to hardware and software infrastructure required for development of high quality startups. The infrastructure facilities are also supplemented by funding support through an innovative fund of fund scheme, which will enable the startups to choose an accelerated trajectory of growth and development.

One of the key differences of startup ecosystem in Kerala is the accessibility of startups to the early stage markets. Government of Kerala has taken the lead role of positioning itself as an early market place for the Startups in the State. By allowing State Government Departments to procure from startups products upto Rs. 20 lakhs without normal tendering process, the State Government has given its intent to embrace innovation and commitment to support high quality startups.

As technology is re-defining the ways of industrial and economic growth, the relevance of small and nimble enterprises capable of bringing innovative solutions for societal problems will be on the rise. However, these are normally lead by high intellect and passionate individuals who have little resources to realize their dreams. By providing right support, infrastructure and resources, a vibrant ecosystem of startups can be developed who could exploit burgeoning opportunities emerging out of new wave of technology revolution. Through multiple interventions ranging from evangelization to skill development to incubation support to funding support to market support, Kerala Startup Mission, in association with other stakeholders has taken up this task of nurturing, hand holding and developing high quality startups which would shape of economy of Kerala in the near future and provide a unique brand personality to its industry and enterprises.

Summary

This chapter looked at the need and evolution of startups in Agriculture. As discussed, startups form a small part in the large canvas of entrepreneurships and created with the objective of developing innovative solutions that have potential to create large scale wealth and/or employment. Normally startup leverages on technology means to address problems of the sector to develop interesting solutions. As Agriculture sector needs to enhance its growth, role of startups for addressing problems of sector cannot be overstated. There is an eminent need to combine the knowledge of Agriculture with new age technologies to develop a breed of successful entrepreneurs. As a Top Performing[3] State, Kerala is constantly developing a vibrant ecosystem for nurturing and developing such startups who through their intellectual depth will come out with sustainable solutions to leap frog the development of Agriculture in the State.

3 Kerala has been awarded as TOP PERFORMER in the State Startup Ranking Exercise 2018 conducted by Government of India (startupindia.gov.in)

Chapter 5

Agri-Clinics and Agri-Business Centers

Dr. A.K. Sherif (Professor & Nodal Officer, Training Service Scheme) and **Binsa B.** (Research Scholar)

Department of Agricultural Extension, Kerala Agricultural University, College of Agriculture Vellayani, Thiruvananthapuram-695 522

Agriculture : Facts & Reality in Kerala

Agriculture and allied sectors are considered to be the foundation of the Indian economy. It forms the important source of raw material and demand for many industrial products. A shift from agriculture to agri-business is a crucial pathway to rejuvenate Indian agriculture and to make it more attractive and beneficial business enterprise. Agripreneurship have the potential to contribute to a range of social and economic development, such as employment generation, income generation, poverty reduction and improvements in nutrition, health and overall food security in the national economy. Agripreneurship has potential to generate growth, diversify the income, and provide widespread employment and entrepreneurial opportunities in rural areas.

Shortage of Qualified Personnel

There is a notion that there is dearth of qualified agri professionals to provide technical know- how or involve in the production of quality inputs. However, the reality is that every year about 500 agriculture graduates, 100 diploma students and 5000 VHSE students (agriculture stream) are graduating from agricultural institutions. Out of this about 200 to 300 are finding job opportunities in private and public sectors. It indicates that more than 5000 graduates are unemployed. This reservoir of technically trained agriculture graduates are waiting for opportunities.

Considering the necessities in the development of agriculture sector to serve the farmers in a better way and utilize the available human resources in the country, many new Central sector schemes has been introduced in the recent past. Agri-clinic and agri-business center (ACABC) is one of the flagship programmes of the Ministry of Agriculture, Govt. of India that was implemented in the country since 2002. This programme had undergone changes and modification, the Government of India has revised the on-going Central Sector Scheme "Establishment of Agri Clinics and Agri Business Centers (ACABC)". The details are available in http://www.agriclinics.net/scheme-training.htm).

Aims of the programme

The scheme now inter alia aims to supplement efforts of public extension by necessarily providing extension and other services to the farmers on payment basis or free of cost as per business model of agricultural entrepreneur (agri-preneurs), local needs and affordability of target group of farmers.

Objectives of the programme

- To supplement efforts of public extension by necessarily providing extension and other services to the farmers.
- To support agricultural development.
- To create gainful self-employment opportunities to unemployed agricultural professionals.

Under this programme, an agriculture professionals trained under this programme could take up agri-clinic or agri-business to supplement the public extension activities.

Agri-Clinics: Agri-Clinics are envisaged to provide expert advice and services to farmers which would enhance productivity of crops/ animals and ensure increased income to farmers. Agri-preneurs can take up the following activities like:

- Soil health testing, cropping practices, plant protection,
- Crop insurance, post-harvest technology
- Clinical services for animals, feed and fodder management
- Prices of various crops in the market, etc.

Agri-Business Centres: They are commercial units of agri-ventures established by trained agriculture professionals for income generation and entrepreneurship development.Agri-preneurs can take up ventures like :

- Maintenance and custom hiring of farm equipment,
- Sale of inputs and other services in agriculture and allied areas,
- Post-harvest management and market linkages.

Eligibility criteria- new guidelines

1. Graduates in Agriculture subjects from State Agri. Universities/ Central Agricultural Universities/Universities recognized by ICAR/UGC.
2. Degree in Agriculture and allied subjects offered by other agencies (Subject to approval of Department of Agriculture & Co-operation, Government of India on recommendation of the State Government).
3. Diploma (with at least 50% marks)/Post Graduate in Agriculture and allied subjects from State Agricultural Universities, State Agriculture and Allied Departments and State Department of Technical Education.
4. Diploma in Agriculture and allied subjects offered by other agencies (subject to approval of Department of Agriculture & Co-operation, Government of India on recommendation of the State Government).
5. Biological Science Graduates with Post Graduation in Agriculture & allied subjects.
6. Degree courses recognized by UGC having more than 60% of the course content in Agriculture and allied subjects.
7. Diploma/Post-Graduate Diploma courses with more than 60% of course content in Agriculture and allied subjects, after B.Sc. with Biological Sciences, from recognized colleges and universities.
8. Agriculture related courses at Intermediate (i.e. plus two) level, with at least 55% marks.

In order to enroll for this programme a candidate have to fill up application in the given format.

(Annexure 1). No upper limit for the age of the candidate. This indicate that retired agri professionals too are eligible for this scheme.

Training and handholding

National Institute of Agricultural Extension Management (MANAGE), Hyderabad, implement and co-ordinate this programme. Nodal Training Institutes (NTIs) in each state will conduct training and guide the professionals for setting up of Agri-Clinics and Agri-Business Centre. There are 56 Nodal Institute in the country and in Kerala, TSS, Vellayani, KAU is one of the Nodal center.

Training programme

- Two months training programme.
- Main aim is to transform a agri graduate into agri- entrepreneurs.
- The training programme is divided into two phases

First Phase

Exposing the trainees into potential agri-ventures (Annexure 2) in their area and providing additional subject matter training to refresh their knowledge and skills. Besides motivation, training will also concentrate on building communication skills, record maintenance, schemes/programmes from where financial assistance is available, etc. For instance, trainees interested in setting up Soil Testing Laboratories, Small Scale Bio-inputs Preparation Units, Custom Hiring of Machineries will be imparted requisite skills and knowledge. The basic idea is to customize the training modules based individual and collective needs of the entire group while adhering to the prescribed syllabi, which is available at www.agriclinics.net/.

Second Phase

Facilitate the trainees to choose a venture and to prepare DPR based on market survey, hands on experience with the support of experienced bankers. Detailed interactions should be arranged with stakeholders, such as farmers, bankers, department officials, input dealers, NGOs working in the area. Identification and analysis of farmers' problems and provision solutions through extension and business advisories are focused upon. Assessing technical feasibility and financial viability of ventures, hands-on interface with the successful entrepreneurs and DPR preparation with assistance by experienced bankers as resource persons are the other crucial parts of this activity.

Handholding : One year

- Interface meetings at State/Institute level are arranged
- Visits to individual enterprises
- Individual progress sheet collected from units
- Monthly progress report to MANAGE

Certificate to successful ventures

- A certificate issued after completion of 2 months training from Nodal center
- Another certificate recognizing that their enterprise is set up under the Scheme.
- This certificate issued after verification & a joint inspection by KAU, PD of ATMA, MANAGE, NABARD, Bank officials
- ***This certificate entitles the agri-preneurs to be recognised by the Development Departments in supplementing the efforts of public extension services.***

Role of ATMA in ACABC

The Extension Reforms Scheme mandates that minimum 10% of resources on extension activities are to be utilized through the non-governmental sector, which also includes Agri-preneurs. ATMAs would be encouraged to implement extension activities through Agri-preneurs. The ventures could also be utilized to provide input services like plant saplings, seeds, and micro-nutrients under on-going flagship schemes implemented by the Department. ATMA officers in development department are part of the monitoring and evaluation process of ACABC units in the state.

Beside this, ACABC is interlinked to Govt. of India Institutions like :

- NSC
- SFCI(State Farming Corporation of India)

- IFFCO
- KRIBHCO(Krashak Bharati Co-operative Ltd. (IFFCO)
- Small Farmers Agri Business Consortium (SFAC)
- State Agro Industries Corporation, etc.

Financial support

1. ACABC Project cost increased from Rs.10.00 lakhs to 20.00 lakhs
2. For group projects from Rs.50.00 lakhs to 100.00 lakhs (group ventures)
3. Relaxation in eligibility criteria-out of 5 people one can be an MBA graduate
4. Subsidy : Back-ended composite subsidy
5. Subsidy for general candidates : 36% of total project outlay
6. Subsidy for women and SC/ST candidates: 44% of total project outlay
7. Working capital component is also eligible for subsidy 3 years lock in period
8. Subsidy Release : Maximum 2 times.

Highlights of the scheme

- Selection of Candidates through screening process
- Uniform Training Schedule for entire country
- Identification of Agri project by Trainees after market survey during training
- Preparation of bankable project by Trainee under the supervision of experts.
- NTI will submit the project reports to Banks
- Hand holding support by MANAGE through NTIs
- Close monitoring of the scheme by a separate cell
- Up front release of subsidy
- Verification of units by committees
- Rewards & Recognition for NTIs and Agripreneurs
- Enhanced loan & subsidy eligibility for extremely successful Agripreneurs.

- Proposal of "One branch one loan"
- Removal of cut off date for subsidy eligibility
- Proposal to cover all activities under CGFT scheme.
- Linkages with ATMA at Block level
- Covering more candidates by relaxing eligibility criteria for ACABC training.
- Certificates for successful Agripreneurs
- Refresher training programme for established Agripreneurs
- Bankers awareness programmes by NABARD
- Monitoring the scheme by separate cell at DOE and MANAGE
- Separate Cell at MANAGE "Centre for Agri-entrepreneurship Development"
- Monitoring the scheme through Consultants
- Separate website for MANAGE and Agri-clinics scheme.

  E-bulletin
- Documentation of success stories
- Encouraging successful entrepreneurs
- Refresher Training programmes to established Agripreneurs
- Settlement of issues by taking up with Ministry of Agriculture and planning commission
- Linkages with ATMA and RKVY
- Business linkages with other Agripreneurs
- Awards for successful Agripreneurs and Nodal Training Institutions
- Monitoring, Sanction of Bank loans and Subsidy releases
- Review meeting with Banks and NABARD :

  For details contact Nodal Officer (ACABC)

  Training Service Scheme

  College of Agriculture, Vellayani (PO)

  Thiruvananthapuram-695522, Kerala

  Ph 94963 13254/0471-2388 061

  Email: acabcgroupkerala@gmail.com

Annexure 1

Application for Agri-clinic and Agri-business Training Programme

Affix a recent Photo here

1. Name:
2. Date of Birth:
3. Sex:
4. Educational Qualification:

   (Attach photocopy of Degree certificate)
5. Permanent Address:

   Village/Town:

   Block:

   District:

   State:

   PIN code:

   Telephone/Mobile:
6. Email ID, if any:
7. Address for communication:
8. Demand draft No:

   (Rs.500/-payable to Manage, Hyderabad, to be remitted after selection)
9. Type of Agri. ventures preferred: Consultancy/Agri business/others
10. Which place you would like to start the above ventures?
11. Which is the nearest BANK to get financial support?
12. Present status : employed/unemployed

Signature

Annexure 2

1. Indicative lists of Agri ventures under ACABC scheme

- Extension consultancy services
- Soil and water quality cum inputs testing laboratories
- Crop protection services, including pest surveillance, diagnostic and control services *(with culture rooms, autoclaves, microscopes, ELISA Kits, etc. for detection of plant pathogens including viruses, fungi, bacteria, nematodes, and insect pests)*
- Micro-propagation including plant tissue culture labs and hardening units;
- Production, maintenance and custom hiring of agricultural implements and machinery including micro-irrigation systems;
- Seed production and processing units;
- Vermiculture units;
- Production of bio-fertilizers, bio-pesticides & other bio-control agents;
- Apiaries (bee-keeping) and honey & bee products' processing units;
- Agricultural insurance services;
- Agri tourism
- Agri journalism–film production, farm publications and exhibitions
- Information technology- establishing of kiosks;
- Poultry and fishery hatcheries;
- Livestock health cover, veterinary dispensaries and services including frozen semen banks and liquid nitrogen supply and artificial insemination;
- Feed production, marketing and testing units;
- Value addition centres;
- Cool chain including cold storage units;
- Post-harvest management centres for sorting, grading,
- Standardization, storage and packaging;
- Metallic and non-metallic storage structures;

- Horticulture clinic, nursery, landscaping, floriculture;
- Sericulture;
- Vegetable production and marketing;
- Retail marketing outlets for processed agri-products;
- Production and marketing of farm inputs & outputs;
- Contract farming;
- Crop production and demonstration;
- Mushroom production;
- Production, processing and marketing of medicinal and aromatic plants;
- Production units like dairy, poultry, piggery, fisheries, sheep rearing, goat rearing, emu rearing, rabbit rearing, etc.

❒❒❒

Chapter 6

Project Formulation

T. Paul Lazarus (Asst. Professor Sr. Scale) and
Ajmal S. (Research Scholar)

Department of Agricultural Economics, Kerala Agricultural University, College of Agriculture Vellayani, Thiruvananthapuram-695 522

Project

Project is an investment activity where we spend capital resources to create a productive asset for realising benefits over time. The project should facilitate analysis in planning, financing, implementation, monitoring, controlling and evaluation.

Important features of a project

It involves more expenditure and more returns.

Its duration is for a long period of time.

For a few initial years, we only incur expenses but do not receive any returns. Only after a few years, we start receiving returns.

The value of money keeps changing over time. Rs. 100/- received in the present year is more valuable than Rs. 100/- received ten years later. So, an ex-ante evaluation of the project requires that the values of costs and returns in various years to be discounted to a common reference year.

Types of agricultural projects

Water resource development projects: e.g. Irrigation, groundwater, land reclamation, salinity prevention, flood control, etc.

Agricultural development projects: e.g. Projects that improve farm economy by using diversified cropping systems as well as farming systems approach.

Agro-industries and commercial development projects: e.g. Input supply, services to farming, processing, storage, market development projects.

Agricultural credit projects: (On-lending projects) e.g. Provide credit to farmers.

Phases in project cycle

1. Conception or Identification,
2. Formulation or Preparation of the project,
3. Appraisal or Analysis,
4. Implementation,
5. Monitoring, and
6. Evaluation.

A. Identification of the Project

In agricultural projects, costs are easier to identify than benefits. The various types of costs involved in the project are:

i) Project Costs: Value of the resources in maintaining and operating the projects.

ii) Associated Costs: Costs that are incurred to produce immediate products and services of the projects for use or sale.

iii) Primary Costs or Direct Costs: These include costs incurred in construction, maintenance and execution of the projects.

iv) Indirect Costs or Secondary Costs: Value of goods and services incurred in providing indirect benefits from the projects such as houses, schools, hospitals, etc.

v) Real Costs and Nominal Costs: Costs at current market prices are nominal costs, whereas if costs are deflated by general price index, these are termed as real costs.

vi) Social Costs: These are technological externalities and technological spill-over accrued to the society due to presence of projects, i.e., pollution problems, health hazards, salinity conditions, etc.

Benefits are split into two: tangible and intangible benefits.

Tangible benefits: Incremental income due to the existence of projects is obtained from the following changes.

Intangible Benefits: These include better income distribution, national integration, better standard of living, etc.

B. Formulation or Preparation of Project

i) Technical Aspects: Soil types, problems associated with different types of soils, potentiality which the soils offer for development, irrigation supply and availability, crops to be grown, availability of desired variety of seeds in required quantities, availability of other complementary inputs as per choice, credit facilities, pests and disease problems, possibility of mechanization, expected yields, storage, processing, marketing facilities, etc.

ii) Financial Aspects: Sources of raising financial assistance and terms and conditions of obtaining finance from the credit agencies, the repayment capacity duly giving allowance for risk and uncertainty. Cashflow chart can be used here.

iii) Commercial Aspects: These aspects focus on the estimation of effective demand, availability of input supplies and arrangements for the output marketing. Market potentiality for the products needs a careful scrutiny.

iv) Managerial Aspects: The managerial skills required; technical skills required. These abilities of the implementing agency needs a closer scrutiny.

v) Organisational Aspects: Organisation refers to the process of putting the priorities in an orderly form. Prepare the organisational hierarchy of the implementing agency. The availability of staff at various cadres, demarcation of authority and linking of authority and responsibility, etc.

vi) Social Aspects: Here customs, culture, traditions and habits, etc., of the beneficiaries are considered. The other relevant implications like the probable changes in the living standards, material welfare, consumption habits, income distribution effects, etc., fall under this.

vii) Economic Aspects: Here we have to examine the benefits, which the project is going to contribute in terms of the utilization of scarce reesources of the nation. Whom the project is going to benefit i.e., to one section of the society or the entire area of the project. The indirect effect like, the income distribution, needs to be assessed. Overall, what one expects of a project is that it should bring in the greatest contribution to the national economy.

C. Project Appraisal

Broadly there are two methods of project appraisal, viz., undiscounted measures and discounted measures.

In the undiscounted measures, payback period, ranking by inspection, proceeds per rupee of outlay, average annual proceeds of rupee outlay, etc., are important.

Under the discounted measures Net Present Worth (NPW), Benefit-Cost Ratio (B-C ratio), Internal Rate of Return (IRR), and Profitability Index are prominent.

Undiscounted Measures

The undiscounted measures are the naive methods of choosing among the alternative projects. The methods listed under these measures often mislead in ranking of the projects and hence, choices go wrong.

Ranking by Inspection: It is based on the size of costs and length of cash-flow stream. Suppose if the two projects are with the same investment and the same net value of production, but with difference in the length of the period, then the project with longer duration is preferred to the one with shorter time period. This leads to bias in the choice obviously due to the absence of more elaborate and appropriate analysis.

Payback Period

It is the length of time required to get back the investment on the project.

The payback period of the project is estimated by using the straight forward formula:

$P=I/E$

Where,

P= Payback period of the project in years,

I = Investment of the project in rupees, and

E = Annual net cash revenue in rupees.

The preference of a particular project is based on the lesser payback period.

Proceeds Per Rupee of Outlay

This is worked out by dividing the total proceeds with the total amount of investment, and a given project is ranked based on the highest magnitude of the parameter.

Average Annual Proceeds of Rupee Outlay

This is another simple choice criterion and in this procedure, total receipts are first divided by the project lifespan and the average proceeds obtained per year are divided by the initial investment on the project. Here too, ranking is given to the projects, based on the highest magnitude of the estimate.

The major drawback with undiscounted measures is that for the same data of the project, we get different rankings, hence, choice process becomes useless. Rankings by these methods are inconsistent and incompatible.

Discounted Measures

Cashflows are yearly net benefits accrued from the project. It they are weighted by discount rate, they become discounted cashflows. These discounted cashflows are the best estimates to decide on the worth of the project. This approach will give the Net Present Worth of the project. The present worth of the costs is subtracted from the present worth of the benefits in order to arrive at the Net Present Worth of the project every year.

Net Present Worth (NPW)

This is simply the present worth of the cashflow stream. Sometimes, it is referred to as Net Present Value (NPV). The choice of discount rate to be used in the measurement of Net Present Worth (NPW) poses many problems. The selection criterion of the projects depends upon the positive value of the net present worth, when discounted at the opportunity cost of the capital. This could be satisfactorily done, provided there is a correct estimate of opportunity cost of capital. NPW is an absolute measure, but not relative.

NPW of the project is estimated using the following equation:

$$NPW = \frac{P_1 + P_2}{(1+i)^{t1}} + \frac{P_3}{(1+i)^{t2}\,(1+i)^{t3}} - C$$

Where,

P_1= Net cashflow in first year,

i = Discount rate,

t = Time period, and

C = Initial cost of the investment

Projects with positive NPW are given weightage in the selection compared to those with negative present values, while zero NPW makes the investor indifferent.

Benefit-Cost Ratio (B-C Ratio)

Here, we compare the present worth of costs with present worth of benefits. While ranking the projects depending upon the B-C ratio, the most common procedure of selecting projects is, to choose the projects, having B-C ratio of more than one, when discounted at opportunity cost of capital. Finally, the given project is opted for implementation, among alternatives based on the highest B-C ratio. Following formula depicts the estimation of B-C ratio:

$$B - C\,Ratio = \frac{\sum_{t=1}^{n} \frac{B_t}{(1+r)^n}}{\sum_{t=1}^{n} \frac{C_t}{(1+r)^n}}$$

Internal Rate of Return (IRR)

It is the discount rate at which the present values of the net cash-flows are just equal to zero, i.e., NPW = zero. When NPW is set equal to zero, the equation is solved for 'i'.

The method of working IRR provides the knowledge of actual rate of return from the different projects. Thus, IRR is known as 'marginal efficiency of capital or yield on the investment'. The IRR must be found out by trial and error with some approximation.

In the working procedure, an arbitrary discount rate is assumed and its corresponding NPW is arrived at. The positive NPW value of the project indicates that IRR is still higher and the next assumed arbitrary IRR value must be comparatively higher than the initial level. This process is continued until NPW becomes negative. Then by interpolation method the exact IRR is found out using the following formula:

$$\begin{pmatrix}\text{Internal}\\\text{rate of}\\\text{return}\end{pmatrix} = \begin{pmatrix}\text{Lower}\\\text{discount}\\\text{rate}\end{pmatrix} + \begin{pmatrix}\text{Difference bet -}\\\text{ween the two}\\\text{discount rates}\end{pmatrix} \left(\frac{\begin{array}{c}\text{Present worth of}\\\text{the cash flow at the}\\\text{lower discount rate}\end{array}}{\begin{array}{c}\text{Absolute difference between the}\\\text{present worths of the cash flow}\\\text{at the two discount rates}\end{array}} \right)$$

D. Implementation

This is the most crucial phase of the project cycle. The secret of successful implementation depends upon the extent of realism put into the plans drawn beforehand. Project implementation can be divided into three different periods, viz., investment period, development period, and full-production period. Investment period may range from few months to few years depending upon the nature of assets to be acquired. Development period too consumes time. Full production period is the time during which the beneficiaries start reaping the benefits of the project. Implementing agency should ensure that the beneficiaries do continue to receive benefits during the entire lifespan of the project.

E. Monitoring

Monitoring is the timely collection and analysis of data on the progress of a project, with the objective of identifying constraints, which impede successful implementation. Monitoring has to be done continuously to offset various shortcomings that crop up from time to time with regard to various aspects of implementation.

F. Evaluation

This is the last phase of the project cycle. It is not confined to the completed project. Evaluation can be done several times during the life of a project. They are: mid-course evaluation, concurrent evaluation and ex-post evaluation.

References

Gittinger, J. Price, 1985. Economic analysis of agricultural projects. Unnumbered series number UNN 76. Washington, DC: World Bank.

Chandra Prasanna, 2009, Projects, Tata McGraw-Hill Education, Delhi.

Reddy S. Subba and P. Raghu Ram, 1996. Agricultural Finance and Management, Oxford & IBH Publishing Co. Pvt. Ltd., Delhi.

□□□

Chapter 7

Stress Management

Dr. B. Seema (Professor & Head, ADG (SZ), and **Navitha Raj** (Research Scholar)

Department of Agricultural Extension, Kerala Agricultural University,College of Agriculture Vellayani, Thiruvananthapuram- 695 522

Stress is defined as "a state of psychological and physiological imbalance resulting from the disparity between situational demand and the individual's ability and motivation to meet those needs." Dr. Hans Selye, one of the leading authorities on the concept of stress, described stress as "the rate of all wear and tear caused by life."

Types of Stress

Stress can be positive or negative: Stress is good when the situation offers an opportunity to a person to gain something. It acts as a motivator for peak performance. Stress is negative, when a person faces social, physical, organizational and emotional problems :

a) **Distress** is a continuous experience of feeling overwhelmed, oppressed, and behind in our responsibilities. Examples of distress include financial difficulties, conflicts in relationships, excessive obligations, managing a chronic illness, or experiencing a trauma. Distress can be classified into two:

   1. Acute stress - Stress which last for only a short period of time

   2. Chronic stress - Stress which last for a longer period of time

b) **Eustress** is the other form of stress that is positive and beneficial. We may feel challenged, but the sources of the stress are opportunities that are meaningful to us. Eustress helps provide us with energy and motivation to meet our responsibilities and achieve our goals. Examples of eustress include graduating from college, getting married, receiving a promotion, or changing jobs. Eustress can be classified into two:

1. Hyper stress - Overloading and over working conditions

2. Hypo stress – When one is unchallenged , it leads to restlessness and uninspired.

Causes of Stress

There are many types of stressors :

- Career Concern: If an employee feels that he is very much behind in the corporate ladder, then he may experience stress. If he seems that there are no opportunities for self-growth, he may experience stress. Hence, unfulfilled career expectations are the significant source of stress.
- Role Ambiguity: It occurs when the person doesn't know what he is supposed to do, on the job. His tasks and responsibilities are not clear. The employee is not sure what he is expected to do. It creates confusion in the minds of the worker and results in stress.
- Rotating Work Shifts: Stress may occur in those individuals who work on different work shifts. Employees may be expected to work on day shift for some days and then on the night shift. That may create problems in adjusting to the shift timings, and it can affect not only personal life but also family life of the employee.
- Role Conflict: It takes place when people have different expectations from the person performing a particular role. It can also occur if the job is not as per expectation, or when a job demands a certain type of behaviour that is against the person's moral values.
- Occupational Demands: Some jobs are more demanding than others. Jobs that involve risk and danger are more stressful. Research findings indicate, job that cause stress needs constant monitoring of equipments and devices, unpleasant physical conditions, making decisions, etc.
- Lack of Participation in Decision-making: Many experienced employees feel that management should consult them on matters affecting their jobs. In reality, the superiors hardly ask the concerned employees before taking a decision. That develops a feeling of being neglected, which may lead to stress.
- Work Overload: Excessive workload leads to stress, as it puts a person under tremendous pressure. Work overload may take two

different forms. Qualitative work overload implies performing a job that is complicated or beyond the employee's capacity. Quantitative work overload is a result of many activities performed in a prescribed time.

- Work Under Load: In this, case, too little work or very easy work is expected on the part of the employee. Doing less work or jobs of routine and simple nature would lead to monotony and boredom, which can lead to stress.
- Poor Working Conditions: Employees may be subject to poor working conditions. It would include bad lighting and ventilation, unhygienic sanitation facilities, excessive noise, and dust, presence of toxic gasses and fumes, inadequate safety measures, etc. All these unpleasant conditions create physiological and psychological imbalance in humans thereby causing stress.
- Lack of Group Cohesiveness: Every group is characterized by its cohesiveness, although they differ widely in its degree. Individuals experience stress when there is no unity among work group members. There are mistrust, jealousy, frequent quarrels, etc., in groups and this lead to stress to employees.
- Interpersonal and Intergroup Conflict: These conflicts take place due to differences in perceptions, attitudes, values and beliefs between two or more individuals and groups. Such conflicts can be a source of stress for group members.
- Organizational Changes: When changes occur, people have to adapt to those changes, and this may cause stress. Stress is higher when changes are significant or unusual like transfer or adoption of new technology.
- Lack of Social Support: When individuals believe that they have the friendship and support of others at work, their ability to cope with the effects of stress increases. If this kind of social support is not available, then an employee experiences more stress.

Potential sources of stress

Environmental factors:

a) Environmental uncertainties of the business cycle.

b) Uncertainties of political system

c) Threats of terrorism in developed and developing nations

d) Innovation that make employee works in similar forms of technical change.

Organizational factors

a) Task demand: it is related to job stress.

b) Role demand: it is related to functioning of an organisation.

c) Interpersonal demand: it is related to the employees demand and satisfaction.

Personal factors

a) Family issue: related to family relationship.

b) Personal economic problems: economic problems of overextended financial resources.

c) Inherited personality: personality problems arise from basic deposition

Approaches to managing stress

1. Knowledge about Stress: In the first stage, an individual should become knowledgeable about stress. He should know about the process and effects of stress. He must find out the major sources of his stress. He must anticipate stressful periods and plan accordingly in advance. He must be honest with himself and decide what he can cope with what he cannot.

2. Physiological Fitness: Exercise in any form can help people in coping with the stress. Non-competitive physical exercise, such as aerobics, walking, jogging, swimming, riding a bicycle, playing softball or tennis has been recommended by physicians as a way to deal with excessive stress levels. There is evidence to suggest that individuals who exercise are much less likely to suffer from certain types of stress related to exercises. With proper exercise, diet control and non-smoking habits, blood pressure and cholesterol become controlled and the body becomes more resistant to pressures. People are more likely to get physically sick or emotionally depressed, if they are overweight or poorly nourished.

3. Time Management: Most of the people are very poor in managing their time. They don't know what must be done and when it would be desirable to do so. The result of poor time management is feeling of work overload, skipped schedules and tension. A well-organised person can often accomplish twice as much as the person who is poorly organised. Therefore, an individual must understand how to manage his time so that he can cope with tensions created by job demands. A few of the well-known time management principles are:
   - Preparing a daily list of activities to be attended to.
   - Prioritizing activities by importance and urgency.
   - Scheduling activities according to the priorities set.
   - Knowing your daily schedule and handling the most demanding parts of a job when you are most alert and productive.
4. Assertiveness: An individual should become assertive. He should not say 'Yes' when he wants to say 'No'. He should start saying 'No' to people or managers who demand too much of his time. Being assertive is an important factor in reducing stress.
5. Social Support Network: Every person should have people to turn to, talk to and rely upon. Good friends become highly supportive during times of stress and crisis. Social network includes friends, family or work colleagues. Expanding your social support system can be a means for tension reduction because friends are there, when needed and provide support to get the person through stressful situations.
6. Readjust Life Goals: Every individual must know what he really wants to do. This should relate to not only the major decisions of the life but to all activities in our life. He must know what is important for him. Because of the severe competition in life to go ahead, most individuals set very high standards and goals for themselves. These high expectations and limited resources to reach such expectations result in stress. Accordingly, every person must read just his goals and make sure he has the ability and resources to reach such goals. Perhaps the goals should be established after the resources have been analysed.
7. Relaxation Techniques: Every individual must teach himself to reduce tension through relaxation techniques, such as Yoga,

meditation, hypnosis and biofeedback. 15-20 minutes a day of deep relaxation releases tension and provides a person with pronounced sense of peacefulness. Deep relaxation conditions will bring significant changes in heart rate, blood pressure and other physiological factors. Yoga is probably the most effective remedy for stress. Studies have revealed that Yoga has cured several stress related diseases.

8. Plan your Life in Advance: So many times, people create situations which induce stress because they neither did not plan nor did a bad job of planning. The traditional Indian attitude of "whatever will be, will be" a way of accepting the unexpected difficulties in life. This attitude may be relevant in those situations over which we do not have any control like death in the family, but for other events in life, it is better to plan in advance, so that we can confront them with confidence when they occur. Individuals may design their own strategies to reduce stress, but it is a must for the organisations to develop programmes that will help the employees in reducing their stress. This will lead to less employee turnover, absenteeism and as a result productivity will improve.

9. Change your attitude: Find other ways to think about stressful situations. "Life is 10% what happens to us, and 90% how we react to it."

10. Be realistic : Set practical goals for dealing with situations and solving problems. Develop realistic expectations of yourself and others.

11. Get organized and take charge : Being unorganized or engaging in poor planning often leads to frustration or crisis situations, which most always leads to feeling stressed - Plan your time, make a schedule, establish your priorities.

12. Take good care of yourself : Eat properly, get regular rest, and keep a routine. Allow yourself to do something you enjoy each day. Paradoxically, the time we need to take care of ourselves the most, when we are stressed, is the time we do it the least.

13. Get regular exercise : Exercising regularly can help relieve some symptoms of depression and stress, and help us to maintain our health.

Six Factors for managing stress

1. Awareness: Date and time, situation, feelings
2. Anticipation
3. Avoidance
4. Appraisal
5. Assertiveness
6. Action

Conclusion

- Identify your stressors
- Know how you react
- Learn techniques to deal with stress
- Integrate stress management into your daily life.

References

Chiesa, A. and Serretti, A., 2009. Mindfulness-based stress reduction for stress management in healthy people: a review and meta-analysis. The journal of alternative and complementary medicine, 15(5): 593-600.

Matteson, M.T. and Ivancevich, J.M., 1987. Controlling work stress: Effective human resource and management strategies. Jossey-Bass.

Quick, J.C., Wright, T.A., Adkins, J.A., Nelson, D.L. and Quick, J.D., 2013. Preventive stress management in organizations. American Psychological Association.

❑❑❑

Chapter 8

Conflict Management

Dr. N. Kishore Kumar (Professor)

Department of Agricultural Extension, Kerala Agricultural University, College of Agriculture Vellayani, Thiruvananthapuram-695 522

Introduction to Conflict

Conflict is largely a perceived phenomenon. It is our perception of the situation that determines, if a conflict exists. Conflict may be healthy or unhealthy; it should not be taken as the opposite order. Though, there is orderliness in conflict, yet it can be disorderly.

Definition

1. Conflict is state of opposition, disagreement or incompatibility between two or more people or group of people.
2. A state of opposition between persons or ideas or interests.
3. A hostile encounter between two or more people.
4. Conflict is usually based upon a difference over goals, objectives of expectations between individuals or groups. Conflict also occurs when two or more people, or groups, compete over limited resources and /or perceived, or actual, incompatible goals.

Types of Conflict

Conflict refers to any situation in which there are incompatible goals, cognitions, or emotions within or between individuals or groups that lead to opposition or antagonistic interaction. The definition recognizes three basic types of conflict:

1. **Goal conflict** is situation in which desired end states or preferred outcomes appear to be incompatible.

2. **Cognitive Conflict** is a situation in which ideas or thoughts are inconsistent.
3. **Affective Conflict** is a situation in which feelings or emotions are incompatible; that is, people literally become angry with one another. Conflict is very common in organizational settings. This is not necessarily a negative feature; the resolution of conflict often leads to constructive problem solving.

Conflict exists in many forms other than the form that can result from competition, and managers should understand the different ways of conflict resolution. Thus, examines conflict from a variety of viewpoints. It first considers the positive and negative aspects of conflict. Next, it discusses the levels of conflict that can occur within organizations. Finally, it identifies some of the basic strategies for managing conflict.

Levels of Conflict

The five levels of conflict are: intra-personal (within an individual), inter-personal (between individuals), intra-group (within a group), inter-group (between groups), and intra-organizational (within organizations):

1. **Intra-personal Conflict :** Intra-personal Conflict, which occurs within an individual, often involves some form of goal conflict or cognitive conflict. Goal conflict exists for individuals when their behaviour will result in outcomes that are mutually exclusive or have compatible elements (both positive and negative outcomes).
   a) **Approach-approach conflict** is a situation in which a person has a choice between two or more alternatives with positive outcomes; for example, a person can choose between two jobs that appear to be equally attractive.
   b) **Avoidance-avoidance conflict** is a situation in which a person must choose between two or more alternatives, and they all have negative outcomes. For example, employees may be threatened with punishment in the form of demotion unless they do something they dislike to spend much time on their job.

c) **Approach-avoidance conflict** is a situation in which a person must decide whether to do something that had both positive and negative outcomes, for example, being offered a good job in a bad location.

2. **Inter-personal Conflict:** Inter-personal conflict involves two or more individuals rather than one individual. Two managers competing for the same promotion, two executives maneuvering for a larger share of corporate capital examples of conflict between individuals are legion and quite familiar.

Reasons

1. Personality differences: Some people have difficulty in getting along with each other. This is purely a psychological problem and it has nothing to do with their job requirements or formal interactions.
2. Perceptions: Varied backgrounds, experiences, education and training result in individuals developing different perceptions of similar realities; the result being an increase in the likelihood of inter-personal conflict.
3. Clashes of values and interests: Conflict that so commonly develops between engineering and manufacturing personnel shows how differences in values might underlie conflict. Members of the engineering department might place a premium on quality, sophisticated design and durability while members of the manufacturing department might value simplicity and low manufacturing costs.
4. Power and status differences: As pointed out by Abraham Zalenznik "Organizations are political structures". They operate by distributing authority and setting a stage for the exercise of power. Similarly, status inconsistencies lead to conflict.
5. Scarce resource: Inter-personal conflict is almost automatic anytime, when there is scarcity. Conflicts over scarce resources are exceedingly common in organizations. Where the scarcity is absolute (the resource level cannot be enhanced), it is very difficult to manage inter-personal conflicts. For example, if three qualified individuals to complete for superior positions in the organization and there is only one such position, interpersonal conflict may develop to an unmanageable level.

3. **Intra-group conflict:** A group experiencing intra-group conflict may eventually resolve it, allowing the group to reach a consensus. Or, the group may not resolve the conflict, and the group discussion may end in disagreement among the members. A study of a large number of groups engaged in business and governmental decision making, tried to identify some the conditions that lead to: (1) the successful resolution of conflict (consensus) or (2) the failure to resolve conflict (disagreement). This study showed that conflict within groups is not a simple, single phenomenon.

Instead, intra-group conflict seems to fall into two distinct categories: (1) substantive conflict and (2) affective conflict.

Substantive conflict refers to conflict based on the nature of the task or to "content" issues. It is associated with intellectual disagreements among the group members. In contrast, affective conflict derives primarily from the group's interpersonal relations. It is associated with emotional responses aroused during inter-personal clashes.

4. **Inter-Group Conflict:** An organization is a collection of individuals and groups. As the situation and requirements demand, the individuals from various others groups will complete. The success of the organization as a whole depends upon the harmonial relations among all inter-dependent groups, even though some inter-group conflicts in organizations is inevitable. The idea is to study inter-group behaviours within an organization so that any conflict can be recognized and dealt with by the management.

5. **Intra-Organizational Conflict:** Four types of intra-organizational conflict exist: (1) vertical conflict (2) horizontal conflict (3) line-staff conflict and (4) role conflict. Although these types of conflict can overlap, especially with role conflict, each has distinctive characteristics :

1. Vertical Conflict: Vertical conflict refers to any conflict between levels in an organization; superior-subordinate conflict is one example. Vertical conflicts usually arise because superiors attempt to control subordinates and subordinates.

2. Horizontal Conflict: Horizontal conflict refers to conflict between employees or departments at the same hierarchical level in an organization.

3. Line-Staff Conflict: Most organizations have staff departments to assist the line departments. The line-staff relationship frequently involves conflict. Staff managers and line managers typically have different personal characteristics. Staff employees tend to have a higher level of education, come from different backgrounds, and are younger than line employees. These different personal characteristics are frequently associated with different values and beliefs, and the surfacing of these different values tends to create conflict.

4. Role Conflict: A role is the cluster of activities that others expect individuals to perform in their position. A role frequently involves conflict.

Managing Conflict

Except in very few situations where the conflict can lead to competition and creativity so that in such situations, the conflict can be encouraged, in all other cases where conflict is destructive in nature, it should be resolved as soon after it has developed as possible, but all efforts should be made to prevent it from developing.

Preventing Conflict

Some of the preventive measures that the management can take place are:

i) **Goal structure**: Goals should be clearly defined and the role and contribution of each unit towards the organizational goal must be clearly identified. All units and the individuals in these units must be aware of the importance of their role and such importance must be fully recognized.

ii) **Reward System**: The compensation system should be such that it does not create individual competition or conflict within the unit. It should be appropriate and proportionate to the group effort and reflect the degree of inter-dependence among units, where necessary.

iii) **Trust and communication**: The greater the trust among the members of unit, the more honest and open the communication among them would be. Individuals and units should be encouraged to communicate openly with each other so that they can all understand each other, understand each other's problems and help each other, when necessary.

iv) **Co-ordination**: Co-ordination is the next step to communication. Properly co-ordinated activities reduces conflict. Wherever there are problems in co-ordination, a special liaison office should be established to assist such co-ordination.

Resolving the Behavioural Conflict

Various researchers have identified five primary strategies for dealing with and reducing the impact of behavioural conflict. Even though different authors have given different terminology to describe these strategies, the basic content and approach of these strategies remain the same. These are:

1. Ignoring the conflict

In certain situations, it may be advisable to take a passive role and avoid it all together. From the manager's point of view, it may be specially necessary when getting involved in a situation, would provoke further controversy or when conflict is so trivial in nature that it would not be worth the manager's time to get involved and try to solve it. It could also be that the conflict is so fundamental to the position of the parties involved that it may be best either to leave it to them to solve it or to let events take their own course. The parties involved in the conflict may themselves prefer to avoid conflict, especially if they are emotionally upset by the tension and frustration created by it. People may intrinsically believe that conflict is fundamentally evil and its final consequences are never good. Thus, people may try to get away from conflict causing situations.

2. Smoothing

Smoothing simply means covering up the conflict by appealing for the need for unity rather than addressing the issue of conflict itself. An individual with internal conflict may try to "count his blessings" and forget about the conflict. If two parties have a conflict within the organization, the supervisor may try to calm things down by being understanding and supportive to both parties and appealing them for co-operation. The supervisor neither ignore or withdraw from the conflict nor does he try to address and solve the conflict but expresses hope that "everything will work out for the best of all." Since the problem is never addressed, the emotions may build up further and suddenly explode. Thus, smoothing provides only a temporary solution and conflict may resurface again in the course of time. Smoothing is more sensitive approach than avoiding as long as the parties agree that not showing conflict has more benefits than showing conflicts, the conflict can be avoided.

3. Compromising

A compromise in the conflict is reached by balancing the demands of the conflicting parties and bargaining in a give and take position to reach a solution. Each party gives up something and also gains something. The technique of conflict resolution is very common in negotiations between the labour unions and management. It has become customary for the union to ask for more than what they are willing to accept and for management to offer less than what they are willing to give in the initial stages. Then through the process of negotiating and bargaining, mostly in the presence of arbitrators, they reach a solution by compromising. This type of compromise is known as integrative bargaining, in which both sides win in a way.

Compromising is a useful technique, particularly when two parties have relatively equal power, thus no party can force its viewpoints on the other and the only solution is to compromise. It is also useful, when there are time constraints. If the problems are complex and many faceted, and the time is limited to solve them, it might be in the interest of conflicting parties to reach a compromise.

4. Forcing

As Webber puts it, "the simplest conceivable resolution is the elimination of the other party-to force opponents to flee and give up the fight-or slays them." This is a technique of domination where the dominator has the power and authority to enforce his own views over the opposing conflicting party. This technique is potentially effective in situations, such as a president of a company firing a manager because he is considered as a trouble-maker and conflict creator. This technique always ends up in one party being a loser and the other party being a clear winner. Many professors in colleges and universities have lost promotions and tenured re-appointments because they could not get along well with their respective chairpersons of the departments and had conflicts with them. This approach causes resentment and hostility and can backfire. Accordingly, management must look for better alternatives, if these become available.

5. Problem solving

This technique involves "confronting the conflict" in order to seek the best solution to the problem. This approach objectively assumes that in all organizations, no matter how well they are managed, there will be differences of opinions which must be resolved through discussions and respect for differing viewpoints. In general, this technique is very useful in resolving conflicts arising out of semantic misunderstandings. It is not so effective in resolving non-communicative types of conflicts, such as those that are based on differing value systems, where it may even intensify differences and disagreements. In the long run, however, it is better to solve conflicts and take such preventive measures that would reduce the likelihood of such conflicts surfacing again.

If there is a single contributory factor that helps in reducing and eliminating negative conflict, it is "trust". Our ability to trust each other has great impact on our working lives, on our family interactions and our achievement of personal and organizational goals. In order to create trust and be trustworthy, it is necessary to avoid aggressive behaviours and at the same time develop supportive behaviours, where people are respected for what they are or what they believe in, and are treated equally without bias or prejudice. In case, a conflict develops at any level, it should be resolved with mutual benefit in mind.

❑❑❑

Chapter 9

Time Management

Dr. Mini B. Nair (Asst. Professor)

Institute of Management in Government (IMG), Thiruvananathapuram-695 033, Kerala

"Time is the scarcest resource and unless it is managed nothing else can be managed."

Peter F. Drucker, the Inventor of Modern Management

We have all heard that time is money, but I believe, it is all that more important for the entrepreneur.

Being an entrepreneur means you are your own manager, as well of others. Being an entrepreneur for comprehensive skills for ensuring success. Some of the skills an entrepreneur is called upon to master are business planning, Human resource management, customer management, sales management and Financial management and hence, the entrepreneur need to have multi-tasking capabilities.

A person who is called upon to do such varied multi-tasking, has to consider time as his prized asset. Money, ideas, talent, support - these can be generated or acquired, but no matter how smart you are, how hard you work, or how determined you are a day goes by in 1,440 minutes.

Effective time management is important for anyone but it's absolutely crucial for entrepreneurs. Because entrepreneurs are ultimately responsible for every aspect of their business, allocating the right amount of time to the right tasks is critical to keeping your business running smoothly. Better time management isn't just about working harder, it's about working smarter.

Dr. Alec Mackenzie, an international expert on time management in his book "The Time Trap" argues that unlike any other resource,'Time' cannot be controlled, replaced or created.One can only manage oneself in relation to it.

Time is also the continuous passage of existence in which events pass from a state of potentiality in the future, through the present, to a state of finality in the past.

And Management is the process of reaching organizational goals by working with and through people and other organizational resources. The 4 basic management functions that make up the management process are Planning, organizing, influencing/ motivating and controlling. Planning and controlling are believed to be the Siamese twins of management.

So, what is Time management? How can we define Time Management?

One popular definition states that time management involves determining one's needs, setting goals to meet needs, and prioritizing and planning to meet goals (Lakein, 1973).

Jones (1998) defines time-management as "the ability to decide what is important in life, both at work and in our home and personal lives." Effective time-management skills can have numerous positive effects on a person's life, both personally and professionally (Jones, 1998). A person or team that is well-organised will be more productive, and have a more positive attitude than a person or team which has poor or no time management abilities (Treacy, 2002). "Good time management provides us with the ability to keep a balance in our lives, or to recognise where the imbalance is." (Jones 1998).

So, Time Management is a process, which consists of the functions of planning, organizing, motivating and controlling the resource Time. Time Management is a skill, an attitude and a technique which one can learn and develop.

Let us go through some Time Management Techniques and tools:

Setting Goals and Objectives

As the captain of the ship leading the organisation towards its vision/objectives, it's your responsibility to set realistic goals and targets in line with the organisational goals both for the team and for yourself. So, you need to learn how to get the most out of the team, however, small or skeletal it is. At the beginning of a period – usually at the start of the year, you should reserve time in your calendar for weekly, monthly, quarterly, semi-annual and annual planning sessions.

Time Planning

Planning is selecting among choices. In order to plan one's use of time, we should first analyse how time is presently being used. This is time analysis.An Activity Log can help you. Purchase a large, week at a glance daily planner.

An Activity Log: (also known as an Activity Diary or a Job Activity Log) is a written record of how you spend your time.

An Activity Log is an eye-opening experience. It will help you to understand whether or not you're doing your most important work during the right time of the day and how you are investing your time.

Activity Logs are also useful for helping you identify non-core activities that can not help you meet important objectives.

To Do List or a "TASK PLAN"

The major reason for poor time management is lack of daily planning. Plan your day well in advance. Prepare a To Do List or a "TASK PLAN". Jot down the important activities that need to be done, high priority work should come on top followed by less important ones.

You should assign blocks of time for working on specific tasks and activities on your list .Congratulate yourself for the work completed.

Prioritize your tasks using the ABCDE method:

A: Tasks you must do—serious consequences, if they don´t get done

B: Tasks you should do—mild consequences, if they don´t get done

C: Tasks you could do—no consequences, if they don´t get done

D: Tasks you delegate

E: Tasks you never do

Here´s the trick: you never do a task 'B' before you have done all the tasks 'Á', and you never do a task C before you have done all the tasks B.

Delegate your workload

It's understandable why most entrepreneurs don't delegate enough-after all, most business owners like being in control. In many cases, it's why they got started with entrepreneurship in the first place. While taking ownership is a requisite trait, it can also be a huge pitfall sometimes.

You must learn the art of delegating more of your work. The right delegating actually takes your role and output to a completely new level. Know yourself well first, and the things that you do well yourself. And then give away those works to others, which you feel you are not as good as than others. As an entrepreneur, the ability to delegate is the true mark of a leader. Delegate jobs, activities and roles which you don't need to do yourselves and those which can be done effectively by others.

Use technology to your advantage

Technology is a helping aid to everyone, including entrepreneurs. Budding Entrepreneurs must acquaint themselves with the latest technologies available which helps to organise and prioritise activities. Software applications, such as Google Calendar or Busy Cal 2 can help you. The ability to harness time-saving technology is the key to lean efficiency.

Don't forget that you can set events on your phone or create a system that helps with customer issues. Use online guidance systems to help reduce the number of customer service calls, allowing your team to be more productive. Technology can work with you. You'll be able to get more done with the limited resources that you have.

Telephone calls

Prepare yourself before you pick up the phone to make a call. Jot down all the important points you want to discuss with the person so that you can be sure, you don't get side-tracked and finally end up making multiple calls to the same person. Train your team members on the same lines.

Emails

Email is number one on the list of items that business owners are losing time on. According to a survey results, business owners are spending 30.9% of their time on email – more than calls (14.2%), direct interaction with employees (24.6%), direct interaction with customers (21%), and direct interaction with vendors (9.4%).

Email and the telephone are a constant source of distraction and temptation. You can reserve 10 to 15 minutes an hour for email and phone calls. Organize email in a systematic way and keep your inbox clean.Create separate folders for each client.

Respond to email with a single sentence. If more is required, dictate the bullet points to your assistant and have them compose reply emails on your behalf. You just review and send.

Communication

Suppose you are interrupted at work by someone at a time, when you can't give him (her) your full attention. You can either make an appointment for a time when you can talk with a clear mind or if you do accept the interruption, give your full attention to the other person. Develop tactfulness, organisational sensitivity, and inter-personal skills. These will help you get more value from all types of conversations, especially brief ones.

During conversations stay alert and watch for body language. Don't try to get an important message across when another person seems pre-occupied. Don't cut important conversations too short: Keeping a conversation's length appropriate to its value gives you a better return on your communication time.

Put your laptop away during meetings. Take notes by hand. You'll connect better with people at the meeting, and you'll have far better recall for what happened. Remember to summarise what you have discussed and decided at all meetings.

Meetings

Meetings fill an increasing number of hours in the workday, and yet most employees consider them as a waste of time.

But there are ways to run effective, efficient meetings that leave your employees feeling energized and excited about their work. Here are some tips:

a) **Make your objective clear.** A meeting must have a specific and defined purpose. Before you send that calendar invite, ask yourself: What do I seek to accomplish? Are you alerting people to a change in management or a shift in strategy? Are you seeking input from others on a problem facing the company? Are you looking to arrive at a decision on a particular matter?

b) **Consider who is invited.** When you're calling a meeting, take time to think about who really needs to be there. If you're announcing a change, invite the people who are affected by it. If you're trying to solve a problem, invite the people who will be good sources of information for a solution. When people feel that what's being discussed isn't relevant to them, or that they lack the skills or expertise to be of assistance, they'll view their attendance at the meeting as a waste of time. Ask yourself ruthlessly: Do all these people really need to attend the meeting? Most of the time, a brief summary by email will work just as well.

c) **Stick to your schedule**. Create an agenda that lays out everything you plan to cover in the meeting, along with a timeline that allots a certain number of minutes to each item, and email it to people in advance. Once you're in the meeting, put that agenda up on a screen or whiteboard for others to see. This keeps people focused.

d) **Come prepared**. Circulate the agenda to all invited executives. The agenda should have a brief on the different issues/ points that will be discussed, list of people attending the meeting and the expected outcome.

e) **Start on time, end on time**. If you have responsibility for running regular meetings and you have a reputation for being someone who starts and ends promptly, you will be amazed, how many of your colleagues will make every effort to attend your meetings. People appreciate it, when you understand that their time is valuable. Another note on time: Do not schedule any meeting to last longer than an hour. Sixty minutes is generally the longest time workers can remain truly engaged.

f) **Summarise.** what has been discussed and decided upon. Always prepare a document (activity list) at the meeting itself which gives the task, action planned, single person who will be responsible for its completion and the timeframe.

g) **Follow up.** It's quite common for people to come away from the same meeting with very different interpretations of what went on. To reduce this risk, email a memo highlighting what was accomplished to all who attended within 24 hours after the meeting. Document the responsibilities given, tasks delegated, and any assigned deadlines. That way, everyone will be on the same page.

h) **Review**. A date/s for reviewing the progress of the action plan agreed upon must be decided in the meeting itself and all attendees should be asked to inform progress of the same by email by this review date/s.

Time Management Matrix

Time management experts like Stephen Covey S R (The Seven Habits of Highly Effective People; Simon & Schuster) have developed a model called a time management matrix. This model enables managers to prioritize their activities and use their time more effectively.With the help of the model, they can evaluate their activities in terms of importance and urgency.

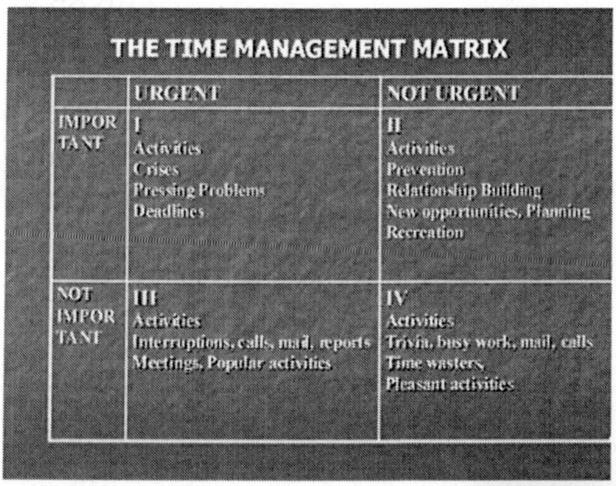

Time Management Strategy

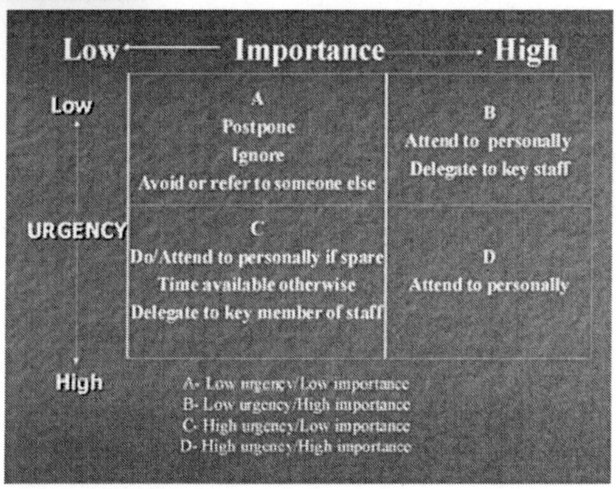

Procrastination

Procrastination is defined as the gap between intention and action. The first step to beat this behaviour is to understand why it happens, is it fear of failure, resistance to change, unpleasant task, do not have the competence to do it,self-doubt, against your principles, wanting to do it all. Avoidance leads to crisis which ends up in more stress and poor outcome.

You need to conquer procrastination by recognizing it and not giving in to it. Start your task right away, as the job well-started is half done. If you feel overwhelmed, then break your project into smaller tasks. Concentrate on each task and complete it, do not worry about the end result. Set your priorities and get motivated to finish these tasks. You will start to develop good work habits and procrastination will not be a problem any longer.

Please remember that "Perfectionism is the greatest Procrastinator" and you need to guard against that.

Consolidate your time

A solid hour is more valuable than numerous 5 minute stretches (even when they add up to an hour), because dispersed moments don't allow for the quality of thought that complicated work requires. Appointments with yourself are a useful (sometimes the only) way to consistently make significant progress on important assignments.

Pomodoro technique,works on the concept of micro breaks.Work/ Focus/Go for 30 minutes on-5 minutes off, 30 minutes on-5 minutes off, 30 minutes on-5 minutes off (this way you can your improve mental agility through short breaks thus, maximizing your performance and increasing your productivity).

Organize your work table/area

De-clutter your world- A desk or work counter covered with papers and other clutter can create an unintended stress. Take a few minutes to scrapit, file it, or delegate it to the right person. A cluttered work space is an indication of a confused mind.

Keep a notepad and a pen handy. Do not write contact numbers or email ids on loose papers. You will waste half of your time searching them.

Files must be kept at their respective drawers. Staple important documents and put them in a proper folder. Learn to be a little more organized. It will save your time which goes on unnecessary searching.

Keep stationery items and your personal belongings like cell phone, car keys, wallet at their proper places.

Conclusion

The business becomes successful, when you transform yourself into a professional manager.The more success you earn as an entrepreneur, the more valuable your time becomes.Awareness or knowledge on Time Management techniques and tools can help one to manage one's mind, that is, it can help in priorities management. When one develops the skill of prioritization of one's work, he starts effectively managing each and every "NOW." This is Time Management when one is effectively managing one's Time, it leads to life management.

Effective time management provides the right opportunity to young entrepreneurs to reap their cake and eat it too. By meticulously adapting the time management techniques you can reduce stress and avoid burnout while living a happier and more productive life. Better time management isn't just about working harder, it's about working smarter.

Reference

1. Covey, Stephen R., Roger A. Merrill & Rebecca A. Merrill: First Things First, New York: Simon & Schuster, 1994.
2. Mina, Eli: The Business Meetings Sourcebook: A Practical Guide To Better Meetings and Shared Decision-Making. New York: AMACOM Books, 2002.
3. Essence of time management, Michael Le Boeuf, Jaico Publishing House, 1995
4. Managing from the Heart, Arun Wakhlu Response Books, 1999
5. Handbook on Time Management Skills for Public Managers – Center for Good Governance, Hyderabad
6. Time Management, Change Management - Center for Good Governance, Hyderabad

7. Yoo, O.S., Corbett, C.J., &Roels, G. (2016). Optimal time allocation for process improvement for growth-focused entrepreneurs. Manufacturing & Service Operations Management, 18(3): 361–375.

8. Yoo, O.S., Roels, G., Corbett, C.J., (2016). Time-money tradeoff for entrepreneurs: When to hire the first employee. *Manufacturing & Service Operations Management 18*(4): 559–569

9. https:/www.scoro.com/blog/tips-for-effective-meeting-management/

10. www.peterbregman.com

11. https://www.forbes.com/sites/forbesleadershipforum/2014/02/05/seven-steps-to-running-the-most-effective-meeting-possible/#15b7a5567a61

12. Patrick Forsyth, Making Meetings Work, CIPD, Publishing, 1998.

13. A review of the time management literature, Brigitte J.C.Claessens, Wendelien van Eerde and ChristelG.Rutte and Robert A. Roe, https://www.researchgate.net/publication/228664480.

❑❑❑

Chapter 10

Leadership Competencies of An Agripreneur

Dhanusha Balakrishnan and **Preethu K. Paul** (Asst. Professors in Contract)

Department of Agricultural Extension, Kerala Agricultural University, College of Agriculture Vellayani, Thiruvananthapuram-695 522, Kerala

"A leader takes people where they want to go. A great leader takes people where they don't necessarily want to go, but ought to be."

Rosalynn Carter

Leadership is lifting a person's performance to a higher standard, the building of a personality beyond his normal limitations. It is the process of influencing the behaviour of others towards the accomplishment of goals in a given situation. Leadership is the art of motivating a group of people to act towards achieving a common goal.The process of encouraging and helping others to work enthusiastically towards objectives. Alan Keith stated that, "Leadership is ultimately about creating a way for people to contribute to making something extraordinary happen".

The essence of leadership

Leadership refers to ability of one individual to influence others. The influence is exercised to change the behaviour of others. Change of behaviour is caused with an objective of achieving a shared goal. The person influencing others (leader) possesses a set of qualities or characteristics with which he or she to influence others leadership is a group phenomenon. It involves interaction between two or more people.

Importance of Leadership

- Leadership transforms potential into reality.
- Leadership is not mere using people and their potential for realising an organisation's goals.

- It has the ultimate aim of raising the level of human conduct and ethical aspiration of both, the leader and the led.
- The leader should elevate, inspire, and evangelise.

Characteristics of effective leader

1. **Guiding vision**: Effective leaders know what they want to do, and have the strength of character to pursue their objectives in the face of opposition and in spite of failures. The effective leader establishes achievable goals.
2. **Passion:** Effective leaders believe passionately in their goals. They have a positive outlook on who they are, and they love what they do. Their passion for life is a guiding star for others to follow, because they radiate promise.
3. **Integrity**: Because they know who they are, effective leaders are aware of their weaknesses. They only make promises they can follow through on.
4. **Honesty:** Leaders convey an aura of honesty in both, their professional and their personal lives.
5. **Trust:** Effective leaders earn the trust of their followers and act on behalf of their followers.
6. **Curiosity:** Leaders are learners. They wonder about every aspect of their charge. They find out what they need to know in order to pursue their goals.
7. **Risk:** Effective leaders take calculated risks when necessary to achieve their objectives. If a mistake is made, the effective leader will learn from the mistake and use it as an opportunity to explore other avenues.
8. **Listening:** Leaders Listen! This is the most important attribute of all, listen to your followers.
9. **Charisma:** This may be the one attribute that is the most difficult to cultivate. It conveys maturity, respect for your followers, compassion, a fine sense of humour, and a love of humanity. The result is that leaders have the capability to motivate people to excel.
10. **Dedication:** The effective leader is dedicated to his or her charge, and will work assiduously on behalf of those following. The leader gives himself or herself entirely to the task, when it is necessary.

Hence, Good leaders are always accountable to others, are great teachers, ask questions and seek counsel, create a positive-energetic atmosphere, genuinely enjoy responsibility, invest in relationships, solve problems without procrastination, challenge people to think, communicate clear expectations, lead by example, make decisions, make others feel safe to speak up, measure and reward performance, properly allocate and deploy talent and provide continuous feedback.

Seven key leadership competencies:

1. Influencing others
2. Fostering Accountability
3. Building Positive Working Relationship
4. Coaching for Improvement
5. Communicating Effectively
6. Working Effectively
7. Building a High Performing Team

Types of Leadership

There are two types of leadership: Formal and Informal leadership.

Formal leadership: Occurs when a manager leads by exercising formal authority. The exercise of formal authority through assigning duties derives, from the manager's official position within the organisation's hierarchy of authority. Any employee who is assigned a managerial position has the opportunity and responsibility to exercise formal leadership.

Informal leadership: Arises when a person without formal authority is influential.

Skills of Effective Leaders

Building and sustaining relationships, developing and communicating a vision, iinfluencing people, making decisions, oovercoming setbacks and adversity and understanding people's needs are the important skills possessed by effective leaders.

An effective leader is a person who does the following:

1. **Creating an Inspiring Vision of the Future**: In business, a vision is a realistic, convincing and attractive depiction of where you want to be in the future. Vision provides direction, sets priorities, and provides a marker, so that you can tell that you've achieved what you wanted to achieve. To create a vision, leaders focus on an organization's strengths by using tools, such as :

Porter's Five Forces, PEST Analysis, and SWOT Analysis, to analyze their current situation:

a) **Porter's Five Forces**: Porter's Five Forces is a simple tool for understanding the competitiveness of our business environment, and for identifying our strategy's potential profitability. This is useful, because, when we understand the forces in our environment or industry that can affect our profitability, we will be able to adjust our strategy accordingly. The tool was developed by Michael Porter, to analyse an industry's attractiveness and likely profitability. Since its publication in 1979, it has become one of the most popular and highly regarded business strategy tools. Porter recognized that organizations-keep a close watch on their rivals, but he encouraged them to look beyond the actions of their competitors and examine what other factors could impact the business environment. He identified five forces that make up the competitive environment, and which can erode your profitability. These are Competitive Rivalry, Supplier Power, Buyer Power, Threat of Substitution and Threat of New Entry.

b) **PEST Analysis**: PEST Analysis is a simple and widely used tool-analyse the changes in our business environment. This helps to understand the "big picture" forces of change that we are exposed to, and, from this, take advantage of the opportunities that they present. It helps to spot business or personal opportunities, and it gives you advanced warning of significant threats. It reveals the direction of change within the business environment. This helps you to shape what we are doing, so that we work with change, rather than against it. It helps to avoid starting projects that are likely to fail, for reasons beyond your control. It can help you break free of unconscious assumptions when you enter a new country, region, or market; because it helps you develop an objective view of this new environment.

c) **SWOT Analysis**: To Discover New Opportunities, Manage and Eliminate Threats. SWOT Analysis is a useful technique for understanding our Strengths and Weaknesses, and for identifying both the Opportunities open to us and the threats we face. Therefore, leadership is proactive-problem solving, looking ahead, and not being satisfied with things as they are.

Once they have developed their visions, leaders must make them compelling and convincing. A compelling vision is one that people can see, feel, understand, and embrace. Effective leaders provide a rich picture of what the future will look like, when their visions have been realized. They tell inspiring stories, and explain their visions in ways that everyone can relate to.

2. **Motivating and Inspiring People to Engage with that Vision:** A compelling vision provides the foundation for leadership. But its leaders' ability to motivate and inspire people that helps them deliver that vision. When we start a new project, we will probably have lots of enthusiasm for it, so it's often easy to win support for it at the beginning. But it's difficult to find ways to keep our vision inspiring after the initial enthusiasm fades, especially if the team or organization needs to make significant changes in the way that it does things. Leaders recognize this, and they work hard throughout the project to connect their vision with people's individual needs, goals and aspirations. One of the key ways they do this is through Expectancy Theory. Effective leaders link together two different expectation: the expectation that hard work leads to good results and the expectation that good results lead to attractive rewards or incentives. This motivates people to work hard to achieve success, because they expect to enjoy rewards-both intrinsic and extrinsic–as a result.

3. **Managing Delivery of the Vision**: This is the area of leadership that relates to management. Leaders must ensure that the work needed to deliver the vision is properly managed-either by themselves, or by a dedicated manager or team of managers to whom the leader delegates this responsibility-and they need to ensure that their vision is delivered successfully. This helps to ensure that the changes needed to deliver the vision are implemented smoothly and thoroughly, with the support and backing of the people affected.

4. **Coaching and Building a Team to Achieve the Vision**: Individual and team development are important activities that leaders need. To develop a team, leaders must first understand team dynamics. A leader will then ensure that team members have the necessary skills and abilities to do their job and achieve the vision. They do this by giving and receiving feedback regularly, and by training and coaching people to improve individual and team performance. By developing leadership skills within our team, we create an environment where we can continue success in the long-term. And that's a true measure of great leadership.

Leadership failures

Inability to organize details, unwillingness to provide support, fear of competition from followers, lack of imagination, selfishness and disloyalty might be the reasons for leadership failures.

So, as a leader you continually increase your ability to realize the best in yourself and to bring out the best in others.

References

Raymer, S.D., Dobbs, J., Kelley, C.P. and Lindsay, D.R., 2018. Leadership Education and Development: Theory Driven Evolutions. Journal of Leadership Education, 17(2): 138-148.

Shamir, B. and Eilam, S.G. 2018. "What's your story?" A life-stories approach to authentic leadership development. In Leadership Now: Reflections on the Legacy of Boas Shamir Emerald Publishing Limited. pp. 51-76.

Gorton, R.A., Alston, J.A. and Snowden, P. 2018. School leadership & administration: Important concepts, case studies, & simulations. McGraw-Hill Education. 760p.

Eagly, A.H. and Johannesen Schmidt, M.C., 2001. The leadership styles of women and men. Journal of social issues, 57(4): 781-797.

❑❑❑

Chapter 11

Problem Solving and Decision-Making

Greeshma Udayan, Cibin J. Das (Project Associate) and
Hridya S. (Research Scholar)
Department of Agricultural Extension, Kerala Agricultural University, College of Agriculture Vellayani, Thiruvananthapuram-695 522, Kerala

Decision-making is an integral part of modern management. Essentially, Rational or sound decision making is taken as primary function of management. Every manager takes hundreds and hundreds of decisions subconsciously or consciously making it as the key component in the role of a manager. Decisions play important roles as they determine both, organizational and managerial activities. A decision can be defined as a course of action purposely chosen from a set of alternatives to achieve organizational or managerial objectives or goals. Decision-making process is continuous and indispensable component of managing any organization or business activities. Decisions are made to sustain the activities of all business activities and organizational functioning.

Decisions are made at every level of management to ensure organizational or business goals are achieved. Further, the decisions make up one of core functional values that every organization adopts and implements to ensure optimum growth and drivability in terms of services and or products offered.

The decision-making process, in a business context, is a set of steps taken by managers in an enterprise to determine the planned path for business initiatives and to set specific actions in motion. Ideally, business decisions are based on an analysis of objective facts, aided by the use of business intelligence (BI) and analytics tools.

In any business situation, there are multiple directions in which to take a strategy or an initiative. The variety of alternatives to weigh -and the volume of decisions that must be made on an ongoing basis, especially in large organizations-makes the implementation of an effective decision-making process a crucial element of managing successful business operations.

Steps in decision-making:

1. Identify a business problem.
2. Seek information about different possible decisions and their likely effect.
3. Evaluate the alternatives and choose one of them.
4. Implement the decision in business operations.
5. Monitor the situation, gather data about the decision's impact and make changes, if necessary.

Data-driven decision-making

Traditionally, decisions were made by business managers or corporate executives using their intuitive understanding of the situation at hand. However, intuitive decision-making has several drawbacks. For example, a gut-feel approach makes it hard to justify decisions after the fact and bases enterprise decision-making on the experience and accumulated knowledge of individuals, who can be vulnerable to cognitive biases that lead them to make bad decisions. That's why businesses today typically take more systematic and data-driven approaches to the decision-making process. This allows managers and executives to use techniques, such as cost-benefit analysis and predictive modeling to justify their decisions. It also enables lines of business to build process automation protocols that can be applied to new situations as they arise, obviating the need for each one to be handled as a unique decision-making event.

If designed properly, a systematic decision-making model reduces the possibility that the biases and blind spots of individuals will result in sub-optimal decisions. On the other hand, data isn't infallible, which makes observing the business impact of decisions a crucial step in case things go in the wrong direction. The potential for humans to choose the wrong data also highlights the need for monitoring the analytics and decision-making stages, as opposed to blindly going where the data is pointing.

Challenges in the decision-making process

Balancing data-driven and gut-feel approaches to decision-making is a difficult proposition. Managers and executives may be skeptical about relying on data that goes against their intuition in making decisions, or feel that their experience and knowledge is being discounted or ignored completely. As a result, they may push back against the findings of BI and analytics tools during the decision-making process.

Getting everyone on board with business decisions can also be a challenge, particularly if the decision-making process isn't transparent and decisions aren't explained well to affected parties in an organization. That calls for the development of a plan for communicating about decisions internally, plus a change management strategy to deal with the effects of decisions on business operations.

Types of decision-making processes

Even when rules and procedures are set up to make business decision-making more systematic, there sometimes can also still be room for intuition on the part of decision-makers. For example, after gathering data about different alternatives, more than one might seem similarly advantageous, or management might find itself lacking certain information needed to make a decision with full confidence. This is a good use case for incorporating intuitive decision-making into the process.

On the other hand, decisions that happen frequently and have clear optimal outcomes benefit from a structured decision-making process. This approach to business problem-solving uses clearly prescribed steps and, usually, data analytics software to evaluate the available options and arrive at a decision.

Sometimes involving more people in the decision-making process can pay off. This is known as participatory decision-making; in the business world, it involves managers seeking input and feedback on decisions from the workers they oversee. The participatory approach has the potential advantage of generating many ideas for solving a business problem; it also helps to engage employees.

References

Allaire, C. and Marsiske, M., 1999. Everyday cognition: age and intellectual ability correlates. *Psychol. Aging.* 14(4): 627–644.

Mehlhorn, K., Ben, A.N., Dutt, V., and Gonzalez, C., 2014. Observed variability and values matter: Towards a better understanding of information search and decisions from experience. *J. Behav. Decision Making.* 27(1): 328–339.

Rokeach, M., 1973. *The nature of human values.* Free Press, New York. 198p.

□□□

II. Crops Production

Chapter 12

Farm Mechanization

Dr. Hajilal M.S. (Professor)

Department of Agricultural Engineering, Kerala Agricultural University, College of Agriculture Vellayani, Thiruvananthapuram-695 522

Farm mechanization is defined as the use of improved tools, implements, machines and equipment on the farm to attain higher productivity and production from the given land using an existing power source. Mechanization can be done in all the farming activities like ploughing, planting, irrigation and drainage, plant protection, harvesting, post-harvest operations, processing, etc.

1. Ploughing

Primary and secondary tillage operations come under ploughing. The operations performed to open up any cultivable land with a view to prepare a seed bed for growing crops are called primary tillage. Implements used are called primary tillage implements. Mainly used primary tillage implements are mould board plough, disc plough, chisel plough, subsoiler and mole plough. Lighter and finer operations performed on the soil after primary tillage are termed as secondary tillage (Operations may be before and after the seed placement in the soil). The secondary tillage implements in general use are disc harrows (single action, double action and offset), drag harrows (blade harrows, spike tooth harrows, spring tooth harrows), cultivators, puddlers, bund former, weeders, levellers, etc.

Tillage Implements classification

1. **Hand operated tools :** spade, shovel, pick axe, hand hoe, etc.
2. **Animal drawn implements:** cultivator, harrow, etc. (riding type or walking type)

3. **Tractor drawn implements:** mould board plough, disc plough, harrows, subsoiler, etc.

Mould board plough

Disc plough

Chisel plough

Subsoiler

Cultivator

Bund former

Puddler

Leveller

2. Planting

Direct sowing of seeds and transplanting of seedlings raised in nursery can be carried by various equipment like broadcasters, seed drills, planters and transplanters. Small seeds are usually placed in a row with the help of seed drills, whereas big seeds like potato are placed in appropriate depth and spacing by planters. Different types of transplanters like walk behind type and self-propelled type are available for paddy cultivation.

Seed drill

Manual rice transplanter

Riding type transplanter

Rice transplanter

3. Irrigation and drainage

Irrigation and drainage are accomplished with the help of suitable pumpsets. Positive displacement pumps and variable displacement pumps are the main two types of pumps classified according to the head discharge relationship. The other types include mixed flow pumps, propeller pumps, jet pumps, air lift pump, etc. Reciprocating and rotary pumps are positive displacement pumps, whereas centrifugal pump is a variable displacement pump. Different types of centrifugal pumps are commonly used for irrigation purpose.

Modern methods of water conserving irrigation systems like drip and sprinkler are now-a-days common. These systems include pumping systems, control valves, filter unit, fertigation unit, flush valves, main line, branch line, laterals, dripper/ sprinkler units, and controller unit for irrigation automation as their components. Proper design and monitoring is essential to obtain high efficiency with these modern irrigation systems.

Centrifugal Pumps

Jet Pumps

Openwell Submersible Pumps

Compressor Pumps

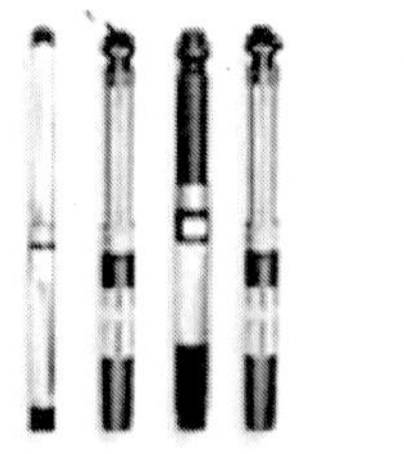

Borewell Submersible Pumps

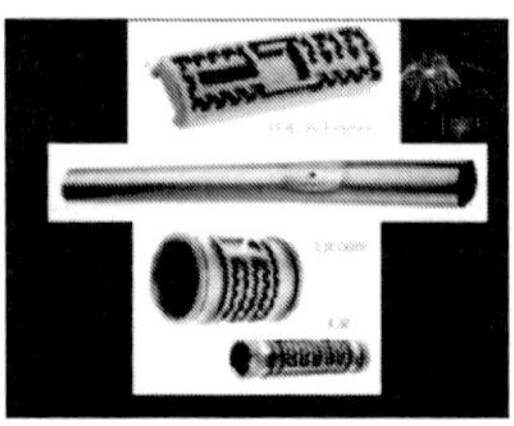

In-line dripper

Emitter

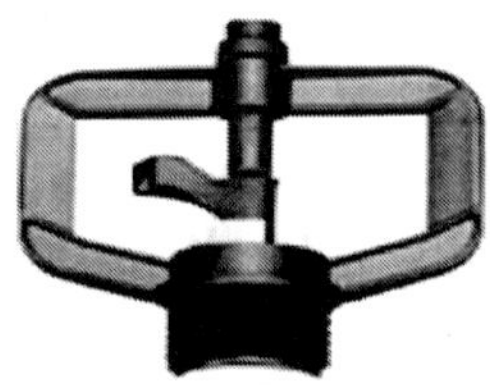

Micro sprinkler

Adjustable sprinkler

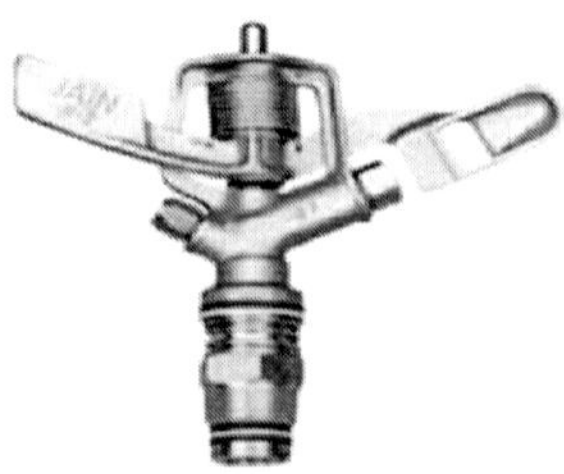

Impact sprinkler

4. Plant protection

The chemicals for protecting the plants from various injuries or organisms need to be applied on plant surfaces in the form of sprays, dusts, mist, etc. Sprayers and dusters are the commonly used plant protection equipment. Sprayer is a machine to apply fluids in the form of droplets, whereas duster is used to apply chemical in dust form. The different types of manually operated sprayers are hand sprayers, foot sprayers, compression sprayers, knapsack sprayers, rocker sprayers/Gator sprayer, stirrup sprayers, etc. The power operated sprayers include hydraulic sprayers, hydro-pneumatic sprayers, air assisted sprayers and foggers/thermal aerosols. The electrostatic sprayers have high application efficiency with lesser use of chemicals but it is very expensive. Aerial sprayers are used for the application of chemicals on very large areas like plantations and recently remote controlled drones are also adopted for the spraying of chemicals.

1. Hand sprayer

2. Hand compression

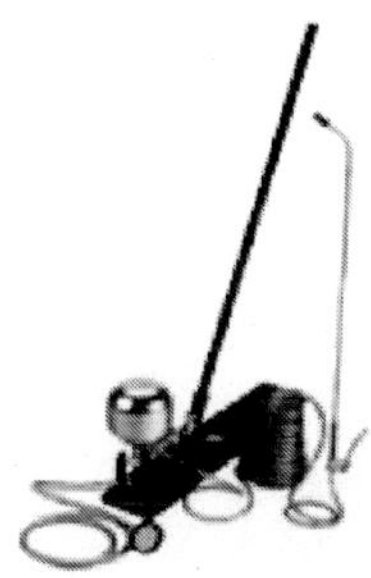

3. Rocker sprayer

4. Power Knapsack sprayer

5. Power Mist blower

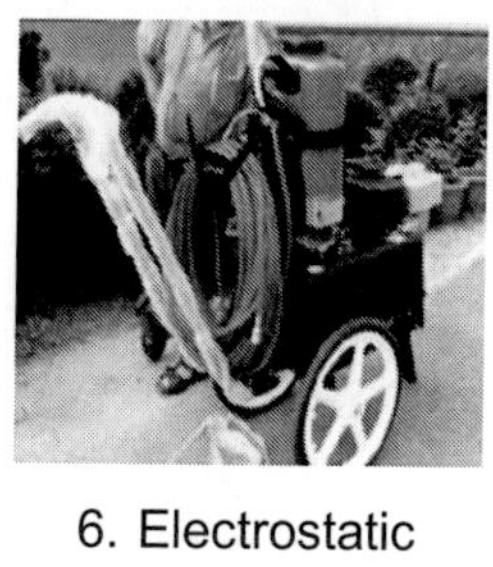

6. Electrostatic sprayer

5. Harvesting

Operation of cutting, picking, plucking, digging or a combination of these operations for removing the crop from under the ground or above the ground or removing the useful part or fruits from the crops is harvesting. It is achieved by: i) slicing action with a sharp smooth edge (knife), ii) tearing action with a rough serrated edge (sickle), iii) high velocity single element impact with sharp or dull edge (brush cutter) and iv) a two element scissor type action (reaper).

Different types of harvesting equipment are: 1) manually operated (sickle, lawn mower) 2) manual-cum-power operated (knapsack type power operated) 3) animal drawn (reaper, mower) 4) engine operated (mower, reaper) 5) power tiller operated (vertical conveyor reaper) 6) Tractor operated (reaper) and 7) self-propelled (combine harvester). Mower is a machine to cut herbage crops not grown in rows and leave them in a swath. Reaper is a machine to cut grain crops cultivated in rows, whereas reaper binder cuts the crops and ties them into neat and uniform bundles. Windrower is a machine to cut crops and deliver them in a uniform manner in a row.

Combine harvester is a machine which performs the functions of a mower/reaper, thresher and winnower. The main functions include feeding the crop to the cutting mechanism (reel), cutting the crop (cutter bar assembly), feeding the cut crops to threshing unit(feeding channel), threshing the crop (threshing drum and concave unit), cleaning the grains from the straw (winnower, straw walker and shaker units), conveying the grains to outlet(elevator), collecting the grains in a container (grain container) and packing of grains (packing unit).

1.Lawn mower

2.Vertical conveyor reaper

3.Combine harvester-paddy

4.Combine harvester-corn

6. Postharvest and Processing

All the operations like threshing, winnowing, dehusking, etc. are considered as post-harvest operations. Processing includes the operation that maintains or improves the quality of farm products or changes their forms. It includes the operations like chaff cutting, grain grinding, cane crushing, grain drying, milk processing, rice processing, seed processing, processing of tea, coffee, rubber, oil palm etc.

1.Manual thresher

2.Paddy winnower

3. Coconut dehusker

4.Groundnut decorticator

5.Tapioca chips maker

References

1. Kumar, S., Kumar, V. and Sahu, R.K., 2016. Fundamentals of Agricultural Engineering, Kalyani Publishers, New Delhi.
2. Ojha, T.P. and Michael, A.M., 2014. Principles of Agricultural Engineering, Vol.I. Jain Brothers, New Delhi.

❑❑❑

Chapter 13

Guidelines for Farming on House Terrace

Dr. V.B. Padmanabhan (Professor & Head and ADE (SZ))

Department of Agricultural Extension, Kerala Agricultural University, College of Agriculture Vellayani, Thiruvananthapuram-695 522

Farming on house terrace is strictly organic farming. Since available land for cultivation is shrinking drastically, especially in urban areas, it is high time that alternative measures like farming on house terrace are to be explored. Farming on house terrace including vegetable cultivation, poultry rearing, azolla cultivation and vermi-composting facilitate better time and space utilization, disposal of household organic garbage, proper harvesting of sunlight, reduced incidence of pests and diseases, reduction of family expenditure, nutritional security due to physical and economic access to vegetables and eggs and above all the better availability of fresh, hygienic, safe and eco-friendly vegetables and eggs to the urban families.

Vegetable Cultivation in Plastic Sacs

a) Plastic sacs are better than earthen flower pots. Among plastic sacs, fertilizer sacs are better.

b) Seventy five per cent of the volume of the plastic sac is to be filled with potting mixture which is prepared by mixing thoroughly soil, sand and powdered cowdung in the ratio 2:1:1.

c) The top edge of the sac is to be folded to have a thick border facilitating irrigation and inter-cultural operations.

d) The sacs are to be arranged in line on the terrace ensuring wall underneath. The sacs are to be placed on bricks placed in a triangular format.

e) Photo insensitive crops and bushy varieties are preferred to photo sensitive crops and trailing varieties.

f) Sowing or transplanting at the rate of two per hill and two hills per sac is advantageous.

g) Transplantation shock to seedlings can be reduced by providing shade by unfolding the edge of the sac and putting a handful of azolla at the base of the seedling.

h) Organic manures are to be applied in sacs at weekly interval.

i) Adequate quantity of water has to be used for irrigation, both in the morning and evening so that excess water will not drip out and plants will not wither.

j) Daily observation and mechanical methods are to be followed for pest control. If situation warrants for the application of pesticides, organic pesticides are to be used.

k) Crop rotation including a pulse crop has to be followed in each sac.

l) After each crop, it is better to change the position of bricks and sacs a little so that easy cleaning of the terrace is facilitated and permanent wetting of any portion of the terrace is avoided.

Poultry rearing under deep litter system

a) Deep litter system has to be followed for rearing poultry on terrace. Approximately, 2 ½ sq. feet area has to be provided for each layer in the netted cage.

b) It is better to avoid the cock and rear the layers alone.

c) The floor of the cage should be without any gap and saw dust has to be used as a good absorbent litter in the cage. The floor should be a raised one leaving space between the terrace and the floor.

d) The cage has to be kept in a shaded corner of the terrace. On the roof of the cage, thick cardboard has to be fitted to reduce the temperature inside the cage. Over the cardboard, polythene sheet has to be spread and fitted so that splashes of rain water will not damage the cardboard.

e) A plastic sheet has to be spread above the cage and tied tightly to poles at distance to provide shade and air column and to protect the cage from rains.

f) A handful of lime has to be dusted over the litter at weekly interval and raked to avoid caking and to act as a disinfectant.

g) Periodic vaccination has to be done to the layers.

h) Plenty of hygienic water and a variety of feed are to be provided in the cage. Low cost and locally available vessels are to be used for feeding which are to be hung in the cage to avoid toppling by the layers. A round plastic plate has to be kept under the water vessel to avoid wetting of the litter.

i) Fresh broken shells are to be kept in separate container as feed for providing calcium to the layers. Powdered dry fish has to be fed to supply protein to the layers.

j) Garlic and onion are to be fed for three consecutive days at monthly interval to avoid infestation by worms and to maintain the general health of the layers. Turmeric powder can be occasionally mixed with feed to act as an antiseptic.

k) Crop residues, fallen leaves, weeds and leaves of chekurmanis, drumstick, papaya, etc. have to be fed to the layers. Fresh azolla can be washed and fed so that the birds will lay 'organic eggs'.

l) Eggs are to be taken out of the cage as early as possible.

m) The litter with the excreta of layers can be occasionally removed from the cage, heaped in a convenient place for a few weeks and then applied as organic manure to vegetable crops.

Azolla cultivation in artificial tanks

Azolla, a floating fern which harbours the nitrogen-fixing blue-green algae can be cultivated in shallow tanks made out of thick polythene sheet in the partially shaded corners of the terrace. Bricks and wood pieces can be used on sides of tank to have a depth of 10 cm on the terrace. Red soil has to be spread with uniform thickness at the bottom of the tank at the rate of 7 kg per sq. metre of area. Fresh cow dung at the rate of 2.5 kg per sq. metre has to be made into a slurry and poured uniformly on the red soil in the tank. Rock phosphate at the rate of 15 gm per sq. metre area has to be given along with cowdung slurry. Water has to be added to the tank to a depth of 8 cm. Healthy azolla at the rate of 250 to 500 gm per sq. metre area has to be spread uniformly in the tank. Azolla starts multiplying after a period of one week.

a) Approximately 50 per cent shade is required for the healthy growth of azolla in tanks on terrace.

b) Polythene sheet of thickness of minimum 150 gauge is conducive for making azolla tanks.

c) Once started multiplying, 250 to 450 gm of azolla can be harvested daily from one sq. metre area.

d) The harvested azolla can be used as organic manure for vegetable crops and as feed for poultry.

e) Removal of a little quantity of water from the azolla tank and addition of fresh water, addition of cowdung slurry and phosphorus at the rate of 0.5 kg and 10 gm, respectively per sq. metre area of tank are to be done at weekly interval.

f) Removal of one-fifth of soil and addition of same quantity of fresh soil to the tank has to be done at monthly interval.

g) The whole tank has to be replaced with new tank once in six months.

h) The production cost of azolla comes to about 50 paise per kg.

i) Mosquito will not grow in tanks, where azolla is grown.

Vermicomposting using household garbage

A unit of vermicompost can be successfully maintained on the terrace utilizing the household garbage as feed for the worms. The plastic container for compost making can be kept on bricks encircled with water to protect it from attack of ants.

The top of plastic container can be covered with GI net to protect it from the attack of rats. A wet gunny sac can be spread over the GI net to provide a cool environment and partial darkness to the worms:

a) The household organic garbages should not be kept exposed before it is put into the vermicompost container, as flies may lay eggs on the exposed garbage which may hatch out into maggots in the vermicompost.

b) When vegetable wastes are continuously put for vermi-composting, dried leaves or pieces of waste paper are to be added to reduce the moisture content.

c) The organic wastes will be devoid of foul smell, when worms act on it.

d) Semi-decayed cowdung is found to enhance the multiplication of worms.

e) The gunny sac on the GI net has to be moistened in the evening everyday.

f) The compost will be ready within a period of one month. As an organic manure, it is better to use the vermicompost as fresh as possible.

References

Padmanabhan, V.B., 2015. Guidelines for urban organic farming on house terrace. In: *Proceedings of Kerala Environment Congress*, 6–8 May, 2015, Kottayam. Centre for Environment and Development, Thiruvananthapuram. Pp. 195-198.

Padmanabhan, V.B. and Swadija, O.K., 2003. Promotion of terrace farming for sustainable agriculture in homesteads- An integrated approach. *National workshop on Homestead farming*, March 6-7, 2003. Farming System Research Station, Kottarakkara. p. 53.

Padmanabhan, V.B. and Swadija, O.K., 2015. Promotion of organic farming in homesteads and urban house terraces- A step to attain nutritional security. In: Oommen, O.V. and Laladhas, K.P. (eds.). *Bio cultural Heritage and Sustainability*. Kerala State Biodiversity Board, Thiruvananthapuram. 240p.

□□□

Chapter 14

Hi-Tech Agriculture

Dr. T. Sajitha Rani (Professor and Head) and
Shilpa E.N. (Research Scholar)
Instructional Farm, Kerala Agricultural University, College of Agriculture
Vellayani, Thiruvananthapuram-695 522

Hi-Tech farming is a knowledge-based agricultural system makes use of technology to increase the value and quality of the farmers produce. It uses computer technology and automation to manage processes, such as watering, fertilizing and pest control to ensure high quantity and high quality outputs.

Hi-tech farming usually takes place in or around densely populated urban areas where land is limited and costly. It is practiced extensively in Europe, North America, and countries, such as Japan and Singapore in Asia.

Hi-tech farming requires high capital inputs. Initial set-up cost of hi-tech farms can be as much as a few million dollars. Hi-tech farming mainly refers to agricultural methods using the latest technology, like hydroponics and aeroponics. There are minimal physical inputs to the system, but it is capital intensive since large capital outlay is required to buy specialized equipment, maintenance, training of labour, etc. It is also a commercial farming system, farmers invest in them both for local and export markets, technical know-how is a must, for example, for hydroponics computers are used to monitor, it is labour intensive because of the training of lab technicians, scientists, chemists and farm hands.

It uses hi-tech farming methods and technology to increase yields; it has high output and market value because the so-called hygienic conditions appeal to all health conscious people out there who do not mind paying more. It is dependent on research and development, like growing temperate vegetables in a tropical climate or genetic engineering for developing disease resistant plants with nice colours or transgenic animals. Hi-tech farming includes use of sophisticated farm machinery like tractor, power tiller, reaper, harvester, weeder, decorticator, ridge maker, etc. It also includes micro irrigation system, precision farming like mulching, use of micro sprayer, etc.

Important components of hi-tech agriculture

Poly house farming

Polyhouse farming is a way of protected cultivation in agriculture. The polyethylene plastic is used to cover the structure. It enables to cultivate high value crops (horticulture) in the structure.

Hydroponics

Hydroponics is a subset of hydroculture, is the method of growing plants without soil, using mineral nutrient solutions in a water solvent. Terrestrial plants may be grown with only their roots exposed to the mineral solution, or the roots may be supported by an inert medium, such as perlite or gravel. The nutrients in hydroponics can come from an array of different sources; these can include from fish waste, duck manure, etc.

Aeroponics

Aeroponics is a system, wherein roots are continuously or discontinuously kept in an environment saturated with fine drops (a mist or aerosol) of nutrient solution. The method requires no substrate and growing plants with their roots suspended in a deep air or growth chamber with the roots periodically wetted with a fine mist of atomized nutrients. Excellent aeration is the main advantage of aeroponics.

Drone technology

One of the latest development in agriculture is the increase in the use of small, unmanned aerial vehicle (UAVs), commonly known as drones. Drones are remote controlled aircraft with no human pilot on-board. These have a huge potential in agriculture in supporting evidence-based planning and in spatial data collection. Despite some inherent limitations, these tools and technologies can provide valuable data that can then be used to influence policies and decisions. Drones can be used in six ways, such as soil and field analysis, planting, crop spraying, crop monitoring, irrigation and health assessment.

Precision farming

Precision farming is generally defined as an information and technology based farm management system to identify, analyse and manage variability within fields for optimum profitability, sustainability and protection of the land resource. In this mode of farming, new information technologies can be used to make better decisions about many aspects of crop production.

Technologies for precision farming are

Global Positioning System (GPS): Global Positioning System satellites broadcast signals allow GPS receivers to compute their location. This information is provided in real time, meaning that continuous position information is provided while in motion. Having precise location information at any time allows soil and crop measurements to be mapped.

Remote Sensing: Remote sensing is collection of data from a distance. Data sensors can simply be hand-held devices, mounted on aircraft or satellite-based. Remotely-sensed data provide a tool for evaluating crop health. Plant stress related to moisture, nutrients, compaction, crop diseases and other plant health concerns are often easily detected in overhead images.

Geographic Information Systems (GIS): Geographic Information Systems (GIS) are computer hardware and software that use feature attributes and location data to produce maps. An important function of an agricultural GIS is to store layers of information, such as yields, soil survey maps, remotely sensed data, crop scouting reports and soil nutrient levels.

Robotics

An agricultural robot is a robot deployed for agricultural purposes. The main area of application of robots in agriculture today is at the harvesting stage. Emerging applications of robots or drones in agriculture include weed control, cloud seeding, planting seeds, harvesting, environmental monitoring and soil analysis.

Benefits of hi-tech farming are:

- Higher crop productivity.
- Decreased use of water, fertilizer, and pesticides, which in turn keeps food prices down.
- Reduced impact on natural ecosystems.
- Less runoff of chemicals into rivers and groundwater.
- Increased worker safety.

Disadvantages

It not only requires a sufficient amount of professional manpower, the equipment and processing of the high tech farming is expensive and tedious. The various types of chemicals and pesticides are dangerous to humans.

References

Aggarwal, S., 2004. Principles of remote sensing. In: Sivakumar, M.V.K., Roy, P.S., Harmsen, K., and Saha, S.K. (eds), *Satellite remote sensing and GIS applications in agricultural meteorology*. 7-11 July, 2003, Dehradun, India, pp. 23-38.

Blackmore, B.S., Stout, W., Wang, M., and Runov, B., 2005. Robotic agriculture - the future of agricultural mechanisation? In: Stafford, J. V. (ed.), *5th European Conference on precision agriculture*, 9-12 June, 2005, Netherlands, Wageningen Academic Publishers, pp. 621-628.

Hakkim, A.V.M., Joseph, A.E., Gokul, A.J., and Mufeedha, K., 2016. Precision farming: the future of Indian agriculture. *J. Appl. Biol. Biotechnol.* 4(6): 068-072.

Mogili, U.R. and Deepak, B.B.V.L., 2018. Review on application of drone systems in precision agriculture. *Procedia Comput. Sci.* 133: 502-509.

Sood, K., Singh, S., Rana, R.S., Rana, A., Kalia, V., and Kaushal, A., 2015. Application of GIS in precision agriculture. In: *National seminar on "Precision farming technologies for high Himalayas,* pp. 04-05.

❑❑❑

Chapter 15

Scope and Importance of Commercial Mediculture in Kerala

Dr. A.S. Anilkumar (Professor)

Department of Agronomy, Kerala Agricultural University, College of Agriculture Vellayani, Thiruvananthapuram-695 522

Herbal medicines known for their safety, efficacy, cultural acceptability and lesser side effects are extensively used in the developing countries for primary health care. Such medicines are available even for age related disorders, for which no modern medicine is available. We are sitting on a gold mine of well-documented and well practiced knowledge of traditional herbal medicine. But we have not been able to utilize native herbal wealth by promoting its use. Insufficient availability of raw materials is one of the major constraints for development of herbal drugs. Excessive unregulated exploitation of medicinal plants and increasing globalization of phytomedicine markets jeopardize future availability and survival of many species. The progressive degradation and loss of forests and other natural ecosystems exacerbates this problem. There is an urgent need to promote domestication and commercial cultivation of medicinal plants to meet the ever increasing demand of user industries. This can be achieved by developing appropriate techniques for micro-propagation, nursery management for quality planting material production, agro climatic delineation for identification of efficient cropping zones, introduction of compatible medicinal crops as inter-crops in the predominant cropping systems, *in situ* rain water harvest, conservation and efficient utilization for early growth and establishment of medicinal trees, organic mediculture, protocols for sustainable harvests, techniques for terrace farming, etc. On farm conservation, product diversification, development of market infrastructure and empowerment of stakeholders in medicinal plants sector are also inevitable for utilization of native herbal wealth.

There is great scope for domestication and commercial cultivation of medicinal plants in Kerala. According to World Health Organization (WHO), about eighty per cent of the population of the developing countries depends on herbal drugs for their primary health care needs, as these are considered as non-narcotic and free from side effects. There is a wide gap between the requirement and availability of medicinal plants. Earlier, medicinal plants were generally collected from the wild. Forests are no longer considered as perennial source of raw material of medicinal plants due to deforestation and unscientific harvests. Domestication and commercial cultivation are the options available for meeting the ever increasing demand of user industries.

Although, 500 medicinal plants are required by the Ayurvedic medicine manufacturing units in large quantities, agro climatic situations prevailing in Kerala are not suitable for commercial cropping of all the species. The success of commercial mediculture depends mainly on selection of appropriate species.

In our state, the scope for pure cropping of medicinal plants is limited due to population density and intensive cultivation. However, there is ample scope for introduction of medicinal plants into the predominant cropping systems of Kerala as most of the medicinal plants are shade loving.

Insufficient availability of quality planting material is one of the major constraints for extension of area under medicinal plants. Hence, development of appropriate nursery management practices is necessary for the production of quality seedlings and saplings of medicinal plants.

Tillage requirement of medicinal plants vary. Due to labour shortage and high cost of labour, farm mechanization is the need of the hour. Tractors and power tillers can be pressed into service for primary tillage. Secondary tillage operations can be carried out based on the tillage requirement of medicinal plants.

Medicinal plants are like other agricultural crops. If the crop is cultivated without application of manures and fertilizers, the soil will be depleted of plant nutrients and production is adversely affected. Hence, developing appropriate strategies for organic mediculture/ integrated nutrient management programme is essential. Soil moisture regulation is also needed for improving productivity and quality enhancement. Problems are experienced in marketing of medicinal plants. Lack of remunerative price for the produce is one of the factors dissuading farmers to extend the area under commercial mediculture.

Lack of assured market is the only limitation to introduce medicinal plants into the existing cropping systems. Unlike other crops, prescribed species of a number of selected medicinal plants are required in different proportions at a particular time for the preparation of the ayurvedic formulation. So, all the medicinal plants required for a particular formulation should be made available to the manufacturer simultaneously by the grower. A grower should cultivate all the selected medicinal plants based on the composition of a formulation in his land area unlike selecting a single species, which is the practice hither to. In this system of cultivation, there is no need for any buy-back arrangement, as the grower himself can prepare the formulation at the farm level itself. It is expected that the grower will get premium price for the produce, as he has a better bargaining power. A system approach based on the requirement of the industry can ensure assured market for medicinal plants.

❑❑❑

Chapter 16

Plant Propagation Techniques in Horticulture

Dr. Deepa S. Nair (Asst. Professor & Head)

Department of Plantation & Spices, Kerala Agricultural University, College of Agriculture Vellayani, Thiruvananthapuram-695 522

Plant Propagation

Propagation can be defined as the creation of new plants from different plant parts or propagules. This can be broadly classified into sexual propagation that involves seeds as the propagating material and asexual propagation that involves vegetative propagules, for generating new plants.

Sexual propagation: The propagule for sexual propagation is seed. The seeds are formed after fertilization, which follows pollination. Pollination occurs, when the pollen falls on the stigma of the same flower or another flower. After pollination, fertilization happens, which is the union of male and female gametes. On fertilization, the ovules of the flower develops into seed and ovary develops into fruit. The fertilization that occurs is called double fertilization. The union of male and female games results in the formation of embryo. The embryo requires nourishment for its development, which is provided by the endosperm, which is formed by the union of egg cells and diploid polar nuclei. Hence, endosperm is triplod. A seed consists of seed coat, endosperm and the embryo.

The seed formed in plants are of two types; orthodox seed and recalcitrant seed. Orthodox seeds are those which can withstand drying to a low moisture level (e.g. seeds of legumes, grasses, etc.). Recalcitrant seeds are those which lose their viability, when dried below a particular moisture level (e.g. seeds of jack, rubber, mango, etc.). In between the two, are the intermediate types. The moisture content of the seed determines its viability. Orthodox seeds, once dried can be stored for long without losing its viability. But recalcitrant seeds cannot be dried and stored. They lose their viability on drying. They have to be sown immediately after extraction for successful germiators.

Seed propagated plants have higher longevity than vegetatively propagated ones. The progeny from the seed need not be true to type. It may show variation from the parent plants.

Asexual propagation

Asexual propagation utilizes propagules, extracted from parent plant like, suckers, stolons, bulbs, tubers, cuttings, etc. The vegetative propagation techniques that requires skill includes budding, grafting, layering and propagation *via* plant tissue culture. Asexual propagation produces true to type plants to their parent plant, as they are produced from the propagules extracted from the parent plant. Hence, this can also be called clonal propagation.

Asexual propagation by specialized structures include suckers, stolons (runners), corms. bulbs, tubers, rhizomes, etc.

Rhizomes are stout, horizontally growing underground stem that put out new shoots and roots. They also are nutrient source that provides proteins and starch for newly developed plants. e.g: ginger, turmeric, canna, etc.

Tubers are fleshy enlarged parts of the stem that forms underground storage structures. They have tiny scale leaves with buds. These buds can form a new plant in asexual propagation and is genetically identical to the parent. e.g. Potato, yams.

Bulbs are asexual propagating structures, which forms the swollen part of the stem within which lie the central shoots of new plants. They are typically underground, and are surrounded by fleshy leaves called scales, that act as nutrient source to the new plant. There are two types of bulbs: a) tunicated bulb (with a dry membrane covering the fleshy scales) and b) scaly bulb. The axillary buds that give rise to new plants are present in the axil of the fleshy scales. e.g. Onion, Garlic.

Corms are solid enlarged fleshy underground stems growing vertically. They store nutrients in their fleshy and solid stem tissue and are surrounded by papery leaves. They are structurally plant stems, with nodes, internodes and buds. These buds may grow up into shoots and adventitious roots develop from their base. e.g. Gladiolus, taro.

Runners are modified stems that grow from existing the stems on the soil surface. They have long and thin internodes. The runners grow in all directions from the parent plant. When used as propagating material, the buds at the nodes produce aerial shoots and adventitious roots. e.g. *Centella, Cynadon,* etc. The runners are also called stolons. Stolons are runners that are produced just below the soil surface.

An offset is a small, complete daughter plant that originates from the axil of the lower leaves of the main shoot. On breaking off from the parent plant, it grows as in independent plant. e.g. Water hyacinth, Cacti, etc.

A sucker is a shoot that arises from an adventitious bud on an underground root or as a lateral branch from the underground axillary bud of an aerial shoot. e.g. Chrysanthemum, mint, banana, etc.

Keikis are specialized structures of propagation of several orchid genera. They are the offshoots which develop on vegetative stems or flower stalks. Examples of plants that use keikis are the *Phalaenopsis, Epidendrum,* and *Dendrobium* genera of orchids.

Cuttings

Cuttings of leaf, stem or root can be used as a vegetative propagating material. A fairly more number of propagules can be obtained from a few number of stock plants.

Leaf Cuttings: In leaf cutting, either the leaf as a whole or a section of the leaf is used as the planting material. The leaf/leaf cutting does not become the part of the new plant that arise from it. It may produce adventitious shoots and roots. The whole leaf with or without petiole can be used as planting material. Matured leaf with 2-3 cm petiole is detached from the parent plant. The lower end of the petiole is inserted into the medium. New plants grow from the base of the petiole. e.g. African violet, episcia, peperomia,etc. In case of plants with sessile leaves, the leaves are cut at the base of the leaf blade along with a lateral bud and are inserted vertically into the medium. The new plant will develop from the bud. In Sanseveria, the leaves are cut length-wise into 10 cm pieces and planted vertically in the medium to produce new plants. In certain plants like *Bryophyllum,* a section of the leaf blade from the margin will initiate new plants. In *Begonia* and *Pepperomia,* the veins on the lower surface of the matured leaf are split and are laid on the medium with the lower surface down.

A small stone is placed above the leaf to keep it pressed in the medium. New plants will develop from each cut.

Leaf-bud cuttings are used for propagating plants like *Clematis, Philodendron, Dracaena, Blackberry, Rhododendron,* etc. These cuttings consist of a leaf blade, petiole, and a short piece of stem with an attached axillary bud. The cutting are planted in the medium with bud inside and leaf exposed.

Stem cuttings

The types of stem cuttings include herbaceous, softwood, semi-hardwood and hardwood cuttings. The polarity of stem cuttings needs to be looked into while planting them in the medium. The cuttings should be placed such that end proximal to the root (of the parent plant) is in the medium. If placed otherwise, rooting will not take place.

Herbaceous cuttings are made from non-woody, soft and succulent herbaceous plants, such as coleus, chrysanthemums, geranium, begonia, etc. A stem of 8-12 cm is cut from the parent plant with intact uppermost leaves. The lower leaves are removed. These cuttings are planted with at least 3 cm below the medium.

Softwood cuttings are prepared from soft, succulent, tip of woody plants. Shoots should be selected such that they snap easily, when bent. The soft shoots require extra care to prevent drying out. Excessive watering may result in rotting of cuttings. These cuttings root in 2 to 5 weeks. e.g. Aster, Dhalia, etc. The complete removal of lower most leaves and partial removal of uppermost leaves adds to the success of propagation.

Semi-hardwood cuttings are prepared from partially mature wood of the current season's growth, after the removal of softwood. Cuttings of 10-15 cm long are prepared with levelled cut at the upper end and a slanting cut at the lower end. The leaves should be partially removed to prevent water loss. e.g., Croton, Hibiscus, etc.

Hardwood cuttings are prepared from mature stems. Only mature stems will root in certain plants. Cuttings can be of 10 to 75 cm. Shoots with long internodes are not preferred for taking the cuttings. Strong and healthy shoots with at least 2 buds should be used. Cutting should be taken with 2 cm below the lower bud and 2 cm above the uppermost bud. Slanting cut should be made at the lower end and levelled cut at the upper end of the cuttings. In case of difficult to root plants, the cutting should be treated with the rooting hormone, auxin. The auxins used for this purpose include Indole Butyric Acid (IBA) and Naphthalene Acetic Acid (NAA). The three types of hardwood cuttings are straight, mallet, and heel. A straight cutting is the most commonly used stem cutting. For the heel cutting, a small section of older wood is included at the base of the cutting. For the mallet cutting, an entire section of older stem wood is included.

Root cuttings: Roots or sections of root are used to initiate new plants. Certain plants initiates new plants only when are still attached to the parent plant. If root cuttings are planted separately, it will not sprout. Only after the new shoot has attained sustainable growth, it has to be detached from the parent plant and planted separately. One such example is curry leaf. Some plants can be propagated from a section of a root. Root cuttings should be taken only from one side of the plant at a time. Next time the roots can be collected from the opposite side. In case of bread fruit tree, roots of 3 cm diameter can be taken from about 100 to 200 cm from the base of the tree. Root cuttings are usually taken from plants during the dormant season, when carbohydrate levels are high. Roots, thus, extracted are prepared into cuttings of 12-15 cm and planted in medium.

Propagation techniques which requires specific skills

This includes grafting, budding and layering, Grafting involves joining together of plant parts by tissue regeneration at the graft union. It involves two or more plants. The upper portion (scion) from one plant is made to grow on the rootstock of another plant. Sometimes, different plants could be grafted on to the rootstock of a single plant. The scion forms the shoot portion and rootstock forms the root portion of the graft by efficient graft union. There are different grafting methods. In some techniques, scions will be attached to the mother plant till graft union is completed and subsequently separated by cutting the scion below and the rootstock above the graft union. This is done in approach grafting or inarching.

In scion detached methods of grafting, the scion of specific diameter and length (depending on the plant species) is excised out from the mother plant and grafted on to the rootstock removing its upper portion. This methods include wedge, saddle, veneer, whip, whip and tongue, softwood and stone (epicotyl) grafting. In order to convert inferior plants to desirable ones, the grafting is done on established plants/trees with scions from plants of desirable traits, which is called top working. To rejuvenate old and worn out trees due to mechanical damage or pest and disease, bridge grafting can be done.

Budding is a form of grafting, wherein the scion is only a single bud rather than a shoot as in grafting. A single bud with or without a portion of the wood is used as the scion, which is inserted into the rootstock until graft union takes place. Different types of budding are: shield budding, T budding, I budding, Patch budding, flute budding, ring budding, forket budding, etc.

Layering is a method of vegetative propagation, wherein root is made to induce on a stem when it is still attached to the parent plant. Once ample roots are formed, they are detached from the parent plant and made to grow as an individual plant. Different types of layering include simple layering, compound (serpentine) layering, tip layering, mount (stool) layering, etc., wherein layers are formed in the ground, in contact with the soil. Another type of layering, is done on the shoot of the plant and not in contact with the ground called airlayering (marcottage or gootee). These layers are usually made on pencil thick shoots. A portion of the bark (possible only in plants with slippable bark), is removed and at the middle of this region a wire or a thread is tied around, so that the bark does not grow and join back. The carbohydrates and endogenous auxins, IAA (Indole Acetic Acid) gets accumulated at this upper margin of the region from which the bark is removed. This promotes rooting. To prevent photo-oxidation of the plant auxins, this region is wrapped with medium using transparent polyethene sheet. When the sufficient numbers of roots are formed, as visible through the polyethene sheet, it is separated from the parent plant and planted as an individual plant.

Micro-propagation is a way of vegetative propagation, where in plant tissue culture techniques are utilized to produce true to type, disease free quality planting materials. In plant tissue culture, very small plant parts like, leaves, nodal segment, cotyledons, pollen, etc. are used to produce plantlets for propagation. These tissues are made to grow *in vitro*, under aseptic condition in artificial culture/growth media under specific conditions of temperature light intensity and humidity. The plants can be mass propagated irrespective of season. A large number of plants can be produced in limited time and space through plant time culture.

References

Hartmann, H.T., Kester, D.E., Davies, F.T. and Geneve, R.L., 1996. *Plant Propagation, Principles and Practices. 6th ed.* Prentice Hall: Upper Saddle River, New Jersey.

Sharma, R.R. and Srivastav, M., 2004. Plant propagation and nursery management. International Book Distributing Company, Lucknow, 488 p.

Singh, J., 2009. Basic Horticulture. Kalyani Publishers, New Delhi, 351 p.

Chapter 17

Quality Planting Material Production in Coconut

Dr. Sreekala G.S. (Asst. Professor)

Department of Plantation & Spices, Kerala Agricultural University, College of Agriculture Vellayani, Thiruvananthapuram-695 522

Coconut is grown globally over an area of 11.988 million ha producing 67042 million nuts (2015-16). The cultivation of coconut extents to 93 countries of the world, of which Asia and Pacific region share a major area and production. India occupies the premium position in coconut in the world accounting for a production of 23904 million nuts (2016-17) from an area of 1.975 million ha. Coconut farming and associated activities provide food and livelihood security to more than 10 million people in India and contributes to about Rs. 20,000 crores to the country's GDP. Kerala occupies the largest area (7.71 laha) and production (7449 million nuts) of coconut in India (CDB, 2018) and the importance of coconut industry in a state like Kerala needs no explanation.

The major constraints in the coconut sector are the decreasing size of operational holdings, the poor economic status of the cultivators, high percentage of senile and unproductive palms, practice of traditional monoculture under rainfed conditions,sudden and unforeseen biotic and abiotic stresses, slow pace of value addition, by-product utilization and slow farm mechanization, particularly in the post-harvest operations. Inadequate availability of quality planting material is another important determining factor in development of a sound coconut industry.Coconut being a perennial crop, which remains in the same area for more than fifty years and a major crop of the state, planting material production assumes utmost importance and care. In the production of planting material for coconut, the production system can be viewed in three aspects.

1. Production of high yielding local cultivars
2. Production of improved selected cultivars/ varieties
3. Production of hybrid coconuts

Establishment of seed garden in different areas are required for the production of quality planting material.Production of high yielding local cultivars and improved selected cultivars/varieties follow the procedure of selection of mother palms, seed nuts and seedlings.

Coconut Nursery

The area where coconut seedlings are produced constitutes the coconut nursery. A nursery is defined as a managed site, designed to produce seedlings grown under favourable conditions until they are ready for planting.

Coconut is propagated by sexual propagation method. The coconut seeds required for raising seedlings should be collected. In the case of coconut, all the required mother tree cannot be planted in the nursery area since it requires a large area for mother block establishment. Hence, selection can be done on the field to identify good high yielding mother plams. This can be done for the production of high yielding local cultivars as well as for common hybrid cultivars where the tall and dwarf parents are commonly available. But for the selected or improved cultivars of specific hybrids, establishment of mother blocks of specific varieties need to be maintained.

1. Selection of site

An appropriate site must be selected for the most effective, efficient, and economical design of a nursery. The nursery site should be preferably with more of sandy soil, near to water source, free from soil pathogens and insects, availability of cheap and skilled labourers and has good access to the main road for easy transportation. In places where sandy soil is not available, the seedbed has to be made up of sand by digging out the required soil, followed by mixing with sand. The design of each nursery should be based on distinct needs, resources, and requirements. Generally, a good nursery should consist of water tank/pond, water pump/pump house, seed and fertilizer store room, implement shed, germination/mother block area; potting/container filling area, seedling raising area, potting shed, office room, propagation structures, compost area, etc. Layout should be in a way that enables operations to flow logically through the nursery so as to save labour and time. Roads and paths within the nursery should be carefully planned. The nursery facilities should be kept clean. Every effort should be made to control weeds in and around the nursery as weeds may host insects and pathogens. The availability of all mineral nutrients is affected by the pH of the growing medium. Hence, growing media should preferably have a pH between 5.5 and 6.5 since maximum availability of nutrients occurs between this pH. For nursery operations like transporting, watering and sales vehicles, machineries and equipments are required. Among them tractor with trolley, wheel barrow, water tanker are necessary.

Open area with some shade trees or protected structure with shade net are the two systems, under which coconut seedlings can be raised. Quality seedlings production starts from the collection of quality seed, nursery establishment and maintenance after its germination. Quality is a qualitative gradable trait which can be measured indirectly by its correlation with growth, productivity, vigour and quality index. While raising the coconut seedlings records regarding the selection of mother plants, numbering of mother plants, management operations undertaken for the mother palms and seedlings, age of mother palms, time of collection, storage period, date of sowing, percentage of germination, culling percentage all has to be recorded. Moreover, sales records, purchase records labour engagement and attendance, pest and disease outbreaks, permanent and temporary stocks and movement register are required to be maintained upto date. There is a need to keep records of expenditure and income, purchase book, sales book, ledger, cash book, dispatch register, etc. Maintaining the records help to provide up-to-date nursery business information, guideline for planning and help to analyze the performance of the nursery activity. From my opinion, a register should also be maintained showing the details of buyers also so that it will be possible to track the records of the plants. An area of 2.5 acres will be required for a nursery, which is expected to raise 10,000 seedlings.

Selection of gardens for mother palan selection:

- Select gardens with a high proportion of heavy bearers and it should not be from very favourable conditions. Garden should be free from the nutritional deficiencies as well as pest and disease incidence.

- Traditional centres are there which are known for collection of good quality seed nuts and seedlings, like Kuttiadi in Kozhikode and Chavakkad in Thrissur districts.

2. Planting Material Production

2.1. Mother Palm Selection

For production of quality planting materials, it is essential to have good quality mother palms of the desired varieties.

Criteria for mother palm selection

- Palm should be a regular bearer with an annual yield of greater than 80 nuts. Palms growing under irrigated conditions should produce at least 120 nuts/palm/year.
- Palms should have reached the full bearing capacity and have consistently higher yields for atleast four years.
- Palms should have straight stout trunk with even growth and closely spaced leaf scars.
- Palms should have short and stout petiole and wide leaf base firmly attached to the stem.
- Palms should have spherical or semi-spherical crown.
- Palms should have more than 30 fully opened leaves and should produce at least 12 inflorescences in a year.
- Palms with short and stout inflorescence stalk with bunches, preferably resting on the leaf petioles of the lower whorl and more number of female flowers (25 or more).
- Husked nuts should weigh not less than 600 g.
- Mean copra content of 150 g per nut or more.
- Bearing nuts of medium size and oblong shape.

Avoid palms which have the following characteristics:

- Palms having long, thin and pendulous inflorescence stalks.
- Palms which produce long, narrow, small sized , irregular shaped and barren nuts.
- Palms which show alternate bearing tendency.

- Palms show shedding of immature nuts in large numbers.
- Palms which are grown under favourable environmental conditions.

Selection of palms in root (wilt) affected areas follow a different pattern :

- In the root wilt disease prevalent tracts, in the midst of heavily diseased palms, high yielding disease free West Coast Tall (WCT), Chowghat Green Dwarf (CGD) and Chowghat Orange Dwarf (COD) palms can be selected as mother palms and open pollinated nuts of those palms can be used for large scale planting material production. Seedlings produced from these palms are expected to be high yielding and disease free.

2.2 Maturity of Seed Nut

In Tall varieties, it takes 11-12 months to become a matured seed nut, whereas in dwarfs, nuts will mature in 10-11 months after emergence of the inflorescence. They produce a resonant and ringing sound, when hit with the harvesting knife or tapped by finger indicating that the husk is dry. Immature nuts will produce a dull sound. Harvesting the bunches with rope is recommended from trees that are very tall and located in places, where the underneath surface is hard.

2.3 Collection of nuts

Harvest seed nuts during the months of December to May in Kerala and February-August in Tamil Nadu to get maximum germination and good quality seedlings.

2.4 Storage of Seed Nuts

Harvested nuts should be stored in shade to prevent drying of nut water. They are stored till their husks become completely dry .For dwarf varieties, storing should be undertaken for only two weeks.In general, seed nuts of tall variety are stored upto two months after harvest and dwarfs are sown within 15 days. For storing, arrange the seed nuts with the stalk-end up over an 8 cm layer of sand in a shed and cover with sand to prevent drying of nut water. Upto five layers of nuts can be arranged one over the other. The nuts can also be stored in plots, provided the soil is sandy and the ground is sufficiently shaded. In the case of nuts harvested in May, heap them in partial shade, till husk is well-dried and then sow them in the nursery. Before sowing, examine seed nuts and discard those without nut water and rotten kernels.

2.5 Raising nursery

Select well-drained, coarse textured soil near dependable water source for irrigation.Raised beds can be taken, if water stagnation is a problem in rainy season. Beds of 1.5m width and convenient length with 75 cm space between beds should be prepared.The seed nuts should be planted in long and narrow beds at a spacing of 30-40 x 30 cm during may- june either vertically or horizontally in 20-25 cm deep trenches. Vertical planting has the advantage of less damage during transit. But when the planting becomes delayed, the nut water goes down considerably and hence, horizontal planting has to be followed then.The nuts collected from particular areas should be kept separately and sown in separate blocks with boards indicating the variety, place of collection, etc. Normally dry coconut fronds and/coir can be utilised as a mulch for the seedbeds (Ghose and Gopalakrishnam, 2013). This can promote early and more efficient germination, good seedling growth and better establishment. A 12% discard of seedlings is acceptable at the pre-nursery seedbed stage (CRIS, 2006; TNAU, 2012).

2.5.1. Raising seedlings in polybags

In polybag seedlings, transplanting shock will be minimized thereby promoting early establishment of transplanted seedlings and seedlings will be with better vigour. Another advantage is that seedlings can be retained longer in the nursery, when conditions for field planting are not favourable. Moreover weeding, watering and elimination of unwanted seedlings as well as seedling selection for planting out to field are more easily accomplished in polybags. For this initially, the nuts are sown very closely in the nursery. Germination of the nuts commences in about three months after sowing and it ends in another two months. Usaully 90-95 per cent of the seed nuts germinate. Seed nuts that donot germinate upto 5 months after sowing are to be removed from the nursery, as they are weak and lack vigour. When they start germinating once in every fifteen days, the sprouted nuts are collected and placed in a black polybag of 500 guage thickness measuring 60 cm x 45 cm filled with potting mixture. The commonly recommended potting media are topsoil mixed with sand in 3:1 ratio or fertile top soil, sand and well-rotten cattle manure/vermicompost in the ratio of 3:1:1. The sprouts are, thus, collected for a period of 5 months from the date of sowing. The disadvantage is on the more expenditure on transportation and seedling cost. (Thomas and Shareefa, 2014). Fertilizers can be applied in the polybags @ 20g ammonium sulphate and 25g muriate of potash per bag after two months of germination and 45g of ammonium sulphate and 45g of muriate of potash per bag after four months of germination. Seedlings are to be watered after application of fertilizers (Nair *et al.*, 2008). Leaf splitting occurs from 4-6 months after polybagging, indicating that the seedlings are ready for field planting.

3. Care and management of nursery

Protective fencing can be provided to nursery, if it is in an open area.For sandy soils, mulching has to be given immediately after the cessation of monsoon. The nursery should be irrigated once in two days during summer months. The nursery beds should be kept free of weeds by periodic weeding. In case of termite attack, the soil in the affected area is removed upto a depth of about 15 cm and the soil should be dusted with carbaryl or chloropyriphos. This should be repeated, if attack persists. Periodic spraying the plants with 1 per cent Bordeaux mixture or any other copper fungicide is recommended to prevent fungal infection (KAU, 2016).

4. Selection of seedlings

Selection of seedlings is an important step for ensuring high yield. Only seedlings with good quality should be selected through rigorous selection:

1. Healthy and robust looking.
2. Early germination, rapid growth and seedling vigour.
3. Early splitting of leaves.
4. Seedlings of 10-12 months old should have 6-8 leaves and nine months old with at least 4 leaves.
5. Collar girth of 10-12 cm

The colour of the petiole and vigour of the seedlings can be used as a selection criterion for dwarfs and hybrids. The Dwarfs should exhibit the petiole colour of the mother palm while, the petiole colour of hybrid seedlings may range from green/brown/intermediate shades of the parents. Seedlings of dwarf varieties can be easily identified by their early germination, short height, short and sturdy leaves with short and narrow leaflets. Dwarf varieties are easily recognized by their colour of petiole and leaves. Seedlings of tall varieties usually grow tall with long leaves and long and broad leaflets. By such rigorous selection, nearly 20-40 per cent of seedlings get rejected.

5. Removal and transportation of seedlings

Seedlings should be removed by lifting with a spade. They should never be lifted from the soil by pulling the leaves or petiole. The seedlings should be planted as early as possible after removal from the nursery. However, under careful storage after removal from the nursery seedlings, can be kept for about four weeks. In such cases, the seedlings should be kept under shade and also watered. Seedlings can be compactly packed and transported. For very long distance transportation, special care should be taken to pack the seedlings in moss/coir pith/other moisture retaining material. Polybag seedlings can be transported as such and planted directly in the field. (Nair *et al.*, 2008). Younger seedlings usually establish more quickly when transplanted than older ones but may notwithstand water-logging in rainy months. Hence, seedlings of two to three years old are generally preferred for planting in bunds of rice fields or similar situations and after planting they have to be held firm by suitable props for about a year till the seedlings strike roots and get established (Thomas and Shareefa, 2014).

Production of coconut hybrids

The usual coconut hybrids are TXD hybrids and DXT hybrids. Apart from this TXT and DXD hybrids can also be produced.

The different steps in the production of coconut hybrids involves:

1. Selection of male and female parents

For TXD mother parent, the selection of mother plants is based on various characteristics relating to yield as discussed earlier for selection of high yielding local cultivars. For DXT hybrids, the dwarf mother palm should show true characters of dwarf parent.

2. Emasculation

The removal of male flower from the inflorescence of the female parent to avoid self-pollination is called emasculation. It is better to do emasculation in the initial few days after opening of the inflorescence. This is done either by removing the individual male flower or by cutting the spikelet above 5 cm from the uppermost female flowers and removing the remaining male flowers by hand.

3. Bagging

A few days before female flowers become receptive, the inflorescence is covered and tied firmly with the pollination bag. Receptivity of a single female flower will last for one to two days.

4. Pollen collection

Collection of pollen from an inflorescence between 6-7 days after opening is recommended. For collection of pollen mature unopened male flowers are collected from fully opened inflorescence. Mature male flower can be easily identified by their bluish green tinge and are ideally collected from middle portion of spikelet. Male flowers are placed between 2 sheets of blotting paper and slightly crushed using a wooden rolling pin to separate the perianth parts taking care not to damage the anthers. The crushed male flowers are dried in incubator at 38-40°C. Fluid bed driers can be used for quick processing of pollen. Using this, it is possible to collect 55-60 g of pollen is about 3-4 hrs. The pollen collected is sieved using 0.2 mm sized sieve. The processed pollen is suitable for immediate use or it can be stored. The collected pollen can be stored in desiccators over $CaCl_2$ for 10-15 days for longer duration of storage. Pollen can be stored in deep freezers for upto 3 months. Pollen can be mixed with neutral talc in a ratio of upto 1 part pollen to 9 part and used for pollination. When stored in deep freezer, the pollen should be allowed to remain at room temperature for a few minute before diluting. One gram of pollen will be sufficient to pollinate 20 to 25 palms.

5. Pollination

The ideal time for assisted pollination is from 8-11 am coinciding with receptivity of the female flowers. For controlled pollination, the first step is bagging of emasculated bunches for the entire period of female flowers and then pollinated with the desired pollen. The receptivity of the female flowers is marked by honey like exudates from the ivory coloured stigma. The inflorescence has to be pollinated repeatedly on the first, third and 5th day starting from the 1st day of opening of female flower. For dwarf plants, since it is having a long female phase, it may be necessary to undertake 2-3 additional pollination on the 7th, 9th and 11th day starting from the 1st day of the female phase. When the stigmas turns brown and black, the female flowers is no longer receptive for pollination. Three days after the receptivity of last female flower, the protecting cloth cover can be removed from pollinating bunch and labelled.

Commercial coconut hybrids

In coconut, many T X D and D X T hybrids have been released for commercial cultivation in view of their higher yield and nature. In most countries, production of D X T hybrids is preferred over T X D hybrids because of the ease with which they are produced, since the climbing of tall palms for the pollination is laborious and time consuming.

D X T hybrid seed production

Here the dwarf palm is the female parent and tall palm, the male parent. In this, natural pollination can be practiced.

For natural pollination, plantations of dwarf should be inter-planted with a single tall cultivar. Such plantation should be established in isolated areas ensuring that no coconut palm exist in the vicinity upto 500 m. to avoid pollen contamination through wind and insect. In this case all male flowers in the inflorescence of female parent are emasculated so that only pollen from tall male parent is available in the garden. Emasculation is done immediately on splitting on spathe by removing all the male flowers leaving only the female flowers for pollination.

For producing D X T hybrids,where the parents are at different location assisted pollination can be done. Pre-potent palms are selected as parents for hybridization. While selecting female parent, following phenotypic character combination are taken into account:

1. Nuts without ridges and having yellow, orange or red colour.
2. Palms with overlapping male & female phase.
3. Small crown and canopy compared to that of tall
4. Narrow stem without any bulges

Hybridization is done carefully using mixed pollen from identified tall palms. Emasculate the inflorescence of the dwarf variety and covered with bags. Conduct assisted pollination using pollen disperser on alternate days for at least 11 days till the last female flower is receptive and fully pollinated. The pollen talc mixture at 1:9 proportion is recommended. Remove bag 7 days after pollination of the last female flower. The nuts should be harvested, before it is sown immediately on the bed without storage. Nursery beds should be mulched or shaded and provided with adequate moisture.

T X D hybrids

There tall palm is the female parent and the dwarf palm is the male parent. In India, the different forms of dwarf variety used as male parent are :

1. Chowghat dwarfs
2. Gangabondam
3. Malayan dwarfs

The tall palm is emasculated and bagged. Pollen is collected from the dwarf palm preferably from the top and middle spikelet of the inflorescence and dusted after dilution with neutral talc in 1:9 proportion. The bags were removed after the female phase is over. A difficulty in T X D hybrid production is the technical difficulty involved in the mass production. The present method requires trained climbers for the purpose of emasculation and hand pollination of all the female flowers of tall mother palms. Hence, the production of T X D hybrids involve considerable labour and expenditure and is time consuming and laborious. Hence, a limited number of hybrid nuts are produced every year.

Some of the T X D hybrids are Kera Sankara (WCT X COD), Chandra Laksha (LO X COD), Laksha Ganga (LO X GB), VHC 1 (ECT X MGD) VHC 2 (ECT XMYD), VHC 3 (ECT X MOD), Ananda Ganga (AO X GB), Kera Ganga (WCT X GB).

Some of D X T hybrids are: Chandra Sankara (COD X WCT), Kalpa Samrudhi (MYD X WCT) and Kalpa Sankara (CGD X WCT)

T X T hybrids

The tall x tall hybrids is produced by the intra-varietal hybridization of the tall variety palms under controlled conditions.

Isolated seed gardens have to be established for the mass scale production of tall x tall hybrids. It should be situated in an area isolated by forest barriers or where no possibility of contamination from foreign pollen exists. In such a garden, the undesirable palms are removed in stages so that the final stand of only highly yielding palm is ensured. The prepotent palms or palms of higher breeding value are identified from among the high yielders in the garden on the basis of seedling programme in the nursery. The palms will serve as the only source of pollen from natural crossing with the emasculated phenotypically superior palms. Because pollination is restricted to a group at high yielding tall palms, it is expected that the resulting tall x tall progenies will give rise to equally promising palms. The tall x tall hybrids may be superior to open pollinated progenies of tall.

Dx D hybrids

The copra quality of dwarf palms are generally inferior. Efforts are also undertaken in various research stations world over for getting precocious and high yielding DXD hybrids.

Selection of hybrid seedlings

Hybrid seedlings are identified in the nursery stage based on characters like :

1. Early germination, petiole colour and seedling vigour.
2. DXT hybrid seedlings can be selected based on exclusive characteristics of the male parent viz.,colour of petiole,length and breadth of leaves and leaflets.
3. One year old seedlings usually show hybrid vigour for collar girth, number of leaves, length and breadth of leaves and leaflets.

Conclusion

Quality planting material is the key to success in the cultivation of any crop and its contribution becomes more important, especially when it comes to the case of perennial crops like plantation crops. If due care is not taken in the production of planting materials, the farmers will face a lot of problems during the course of cultivation incurring heavy loss. Presently, morphological traits are used for varietal identification, seed purity and hybrid assessment while molecular markers are in various stages of development in coconut.

The production of quality planting material depends on those who are involved in the production scenario and thus it becomes the duty of them to offer good quality planting material since it contributes towards high productivity of coconut.

References

CDB, 2018. (Coconut Development Board) State/district/ blockwise area, production, and productivity of coconut. Available: http://coconutboard.nic.in/Statistics.aspx [12 August, 2018]

CRIS [Coconut Research Institute of Sri Lanka], 2006. Nursery management and seedlings selection. Advisory Circular No. A2. Available : www.cri.gov.lk [July 5, 2018]

Ghose, S. and Gopalakrishnan, R., 2013. *Coconut Guide – 2013*. Coconut Development Board. Kochi, Ministry of Agriculture, Government of India.

KAU (Kerala Agricultural University), 2016. Package of Practices Recommendations: Crops (15th Ed.). Kerala Agricultural University, Thrissur, 393p.

Nair, R.V., Jerard, B.A. Samsudeen, K. and Niral, V. 2008. *Coconut Nursery Management*. Central Plantation Crops Research Institute, Kasaragod.

Thomas, R.J. and Shareefa, M., 2014.Advances in planting material production of coconut. In : *Training Manual on Advances in Production of Quality Planting Materials in Plantation Crops*, pp. 1-8.

TNAU. [Tamil Nadu Agricultural University, 2012. Expert system for coconut. Tamil Nadu Agricultural University, Agrictech Portal - Expert System. Available: http://agritech.tnau.ac.in/expert_system/coconut [10 June, 2018]

□□□

Chapter 18

Fodder Production Technology for Kerala Home Gardens

Dr. Usha C. Thomas (Asst. Professor), **Ishrath P. K., Ajmal Fayique C. and Sruthi Liz Thomas** (Research Scholars)

Department of Forage Crops Utilisation, Kerala Agricultural University, College of Agriculture Vellayani, Thiruvananthapuram-695 522

Kerala, the state with highest percentage of crossbred animals in India, having higher potential for milk production, has a low average milk yield of only 7.508 kg milk cow^{-1} day^{-1}. The insufficiency in milk production leads to import of milk from other neighbouring states. Shortage of good quality green fodder is the main reason for low productivity. The green fodder plays a vital role in the milk industry. In Kerala, due to ever-increasing population pressure of human beings, arable land is mainly used for food and cash crops, and the per capita land availability is decreasing day-by-day. Thus, there is little chance of having arable land accessible for fodder crops alone. But Kerala has the lively mixed farming system called home gardens from time immemorial. Home gardens offer an array of opportunities for agrarian sector of Kerala. The components may be food crops, cash crops, animal husbandry activities, etc. Apart from the income generation activities, it offers mental relaxation as well as develops family attachments. So, the inclusion of fodder crops into the existing homestead farming systems has a vital role in Kerala dairy farming sector.

Coconut-based cropping systems are popular in Kerala. This system is inimitable to Kerala and is highly intensive and diverse with a variety of crops and animals immediately surrounding the home. In Kerala, coconut is the main crop in the home gradens and these coconut gardens offer good opportunity for fodder production also. The development of livestock sector has proved to be very important, as it is the major livelihood of over a million households. In Kerala, majority of livestock owning farmers are either small and marginal or even landless. Fodder crops are cultivated on 7,000 ha, which produces green fodder to meet only about 2 per cent of the total dry fodder requirement of the State. The important sources of dry matter available are crop wastes, such as paddy straw, banana pseudostems and leaves, coconut leaves, pineapple waste, jack leaves, cocoa waste, road side grazing, weeds, and other materials which are very poor in nutrients. There is huge scarcity of green fodder in the summer season. Besides these sources, dairy farmers depend on highly priced concentrates for feeding cattle which offset their profit to a considerable extent. Hence, fodder cultivation will commonly find place in the existing homestead farming system. Studies conducted at CPCRI, Kasaragod have shown that growing of crops in the coconut garden has benefitted farmers in terms of increased economy apart from improving the soil health. Growing grass in the inter-space of coconut can increase production of green fodder which can boost the livestock production. The availability of more green fodder will also increase the profit from livestock by reducing the expenditure on costly feed concentrates. This will encourage coconut, based mixed farming system for doubling the income of farmers apart from ecologically sustainable farming model (Thomas *et al.,* 2018).

Perennial fodder grasses like hybrid napier and guinea grass are widely accepted by the dairy farmers all over Kerala, as these grasses are well-adapted to tropical conditions with potential for higher yields per unit area and shade tolerance (Posler *et al.,* 1993).

1. Grasses

Grasses are cultivated mainly for cut-and-carry systems. Most of the fodder crops come under this group. This includes both, annuals and perennials. The grasses can be preserved for future as hay and silage. The important fodder grasses suitable for Kerala are guinea grass, hybrid napier, para grass and congosignal grass.

1.1. Guinea grass *(Panicum maximum)*

Guinea grass is a popular fodder grass of the tropics suited to the agro-climatic conditions of Kerala. It can be profitably grown as a component of agro-forestry systems and comes up well under coconut and other trees. As an excellent fodder, it is much valued for its high productivity, palatability and good persistence. It is a perennial bunch grass, 0.5 to 4.5 m high. The stem is stout to slender, erect or ascending, glabrous or hairy. Leaves are 10 to 100 cm long and 3.5 cm wide. panicle loose and much branched, the lowermost branches being in a distinct whorl. The small seeds are enclosed in smooth glumes. The seeds shatter. The root system is deep, dense and fibrous.

Varieties

Haritha

A semi-perennial developed by mutation breeding from FR 600. It is shy flowering, has glabrous leaves, yields 60-70 t/ha and is suitable for partial shade in coconut gardens. The grass is nutritious, palatable and free from oxalates. It can be cultivated in coastal sandy tracts of Kerala throughout the year.

Marathakam

A high yielding guinea grass variety developed by mutation breeding from FR 600. It is semi-perennial giving a yield of 70-80 t/ ha^{-1} with better fodder quality and tolerant to partial shade. The grass is nutritious, palatable and free from oxalates. It is suitable for cultivation in uplands and homesteads of Kerala throughout the year.

Harithasree

A high yielding guinea grass variety developed by clonal selection from JHGG- 96-3. It is high yielding and having better quality. The green fodder yield recorded is 80-100 t ha^{-1}. The stem is pubescent and leaf is glabrous. It is high tillering with dark green leaves. The leaves are long and soft. The variety is free from pests and diseases. The crude protein content is 8% and crude fibre content is 28%. The grass is nutritious, palatable and free from oxalates. It is suited for cultivation in uplands and homesteads of Kerala.

Other important varieties are Makueni, Riversdale, Hamil, PGG-4 and FR-600. Makueni is a drought resistant cultivar suited to rainfed situations in the state.

Cultivation practices

Climate

Guinea grass thrives well in warm moist climate. It can grow from sea level to 1800 m altitude. It is frost sensitive. It thrives between a temperature range of 15 to 38 °C. The grass tolerates shade and grows under trees and bushes, and is best suitable as an inter-crop in coconut gardens.

Soil

The grass is adapted to a wide range of soils. It usually grows on well-drained light textured soil,preferably sandy loams or loams, but is better suited to medium to highly fertile loams. It cannot tolerate heavy clays or prolonged water-logging.

Season

Under Kerala conditions, the best season of planting is with the onset of south-west monsoon during May-June. As an irrigated crop, planting can be done at any time of the year.

Planting material

Seeds and slips can be used as planting material. Since seed germination is poor vegetative propagation is preferred. To obtain slips for planting, old clumps are uprooted and slips with roots are separated. For planting one hectare, 1.25 lakhs of slips are required. If seeds are used (3 kg ha^{-1}), it should be sown in nursery and the seedlings transplanted in the main field.

Land preparation

The grass requires thorough cultivation to prepare a weed-free seedbed for establishment. For this, two or three ploughings and one levelling are sufficient. In the prepared field, trenches of 10 cm width and 20 cm depth are made. In these trenches, FYM should be applied alongwith phosphorus and potassium fertilizers. Mix with soil and cover the trenches and form ridges of 15 cm height for planting slips.

Spacing

The spacing of 40 cm x 20 cm is followed, when grown as an inter-crop. For a pure crop, a wider spacing of 60 cm x 30 cm is required. Slips are planted on ridges @ three slips per hill.

Nutrient management

In acid soils, application of lime@ 500 kg ha^{-1} in alternate years is desirable. A basal dose of 10 tonnes of FYM, 50 kgP_2O_5 and 50 kg K_2O ha^{-1} (applied in trenches) is recommended. For top dressing,use 200 kg N ha^{-1} in two split doses, the first dose immediately after-first cutting and the second dose during the north-east monsoon period. If irrigation facilities are available, top dressing can be given in more splits. The fertilizer may be applied on either side of the plants, along the row and earthed up.

Irrigation practices

At planting two irrigations are required within seven to ten days for quick establishment. The crop should be subsequently irrigated depending upon the rainfall and soil type. Usually irrigation once in 7-10 days is required. Irrigation with cowshed washing or sewage water within 3-4 days after cutting gives better growth.

Inter-cultivation

The delicate seedlings or newly emerged shoots from slips or cuttings require protection from weeds in the first two months. Two inter-cultivations should be given during this period. Later, inter-cultivation maybe necessary after three or four cuttings.

Harvesting

The crop is ready for harvest, when it reaches 1.5 m height. Cutting at 15 to 20 cm above the ground level is advised. The first cut is usually ready in 9-10 weeks after planting and subsequent cuts are taken at 45 to 60 days intervals. About six to seven harvests can be made in a year. Approximately, 80-100 t ha^{-1} of green fodder is obtained per year.

Inter-cropping

Guinea grass can be grown mixed with leguminous fodder crops, such as cowpea and stylo.

Nutritional value

The grass is nutritious, palatable and free from oxalates. It makes good hay and silage. The crude protein and the crude fibre content of this grass vary from 8 to 14 per cent and 28 to 36 per cent, respectively.

1.2. Hybrid napier *(Pennisetum typhoides x P. purpureum)*

Hybrid Napier produces more tillers and numerous leaves. It grows faster and produces more herbage but the stems are hard and the plants less persistent. The stems are also less fibrous. The tillers are more numerous and grow faster.

Varieties

Suguna

A bajra-napier hybrid developed by crossing Composite 9 and FD 431. It is having high tillering capacity (40 tillers plant^{-1}) with long broad leaves. The leaf sheath is pale green in colour with purple pigmentation. The leaf margin is serrated. The average inter-nodal length is 6.5 cm and leaf stem ratio is 0.82. It has better quality with crude protein content of 9.4% and crude fibre content of 24.5%. The grass is nutritious, palatable and free from oxalates. The hybrid recorded a green fodder yield of 280-300 t ha^{-1}. The yield increase over local check is 36.6%. The variety is free from pests and diseases. It is suited for uplands, homesteads and rice fallows of Kerala.

Supriya

A bajra-napier hybrid developed by crossing TNSC 4 and FD 471. It is high tillering with 35 tillers per plant. The leaf sheath is pale green in colour with long broad leaves. The average inter-nodal length is 10 cm and the leaf stem ratio is 0.76. It has better quality with crude protein content of 9.0% and crude fibre content of 26.0%. The grass is nutritious, palatable and free from oxalates. The hybrid recorded a green fodder yield of 270- 290 t/ha^{-1}. The yield increase over local check is 31.3 %. The variety is free from pests and diseases. It is suited for uplands, homesteads and rice fallows of Kerala.

The popular hybrids are: Pusa Giant Napier, Gajraj, NB-5, NB-6, NB-21 and NB-35

Cultivation practices

Climate

The grass grows throughout the year in the tropics. The optimum temperature is about 31°C. Light showers alternated with bright sunshine is very congenial to the crop. Total water requirement of the grass is about 800-1000 mm.

Soil

Hybrid Pennisetum can grow on a variety of soils. Light loams and sandy soils are preferred to heavy soils. The grass does not thrive well on water-logged and flood prone lands. Phenomenal yields are obtained from very deep fertile soil rich in organic matter. It tolerates pH ranging from 5 to 8. Hybrid Napier requires a deep, thorough weed free and compact seedbed. Three or four ploughings followed by disc harrowing is ideal.

Season

Planting is done with the onset of south-west monsoon.

Planting material

Being a sterile hybrid, the grass is propagated through rooted slips or stem cuttings. Cuttings of moderately mature stem (3 months old) and preferably from the lower two-third of the stem length sprout better than older stems. The cuttings with three nodes are stuck into the soil with the basal end down, either vertically or at an angle to such a depth that two nodes remain within the soil and one above the soil surface. The underground nodes develop roots and shoots while the upper ones develop shoots only.

Spacing

A spacing of 60 cm x 60 cm is recommended for pure crop of Hybrid Napier. In inter-cropping system, spacing is adjusted to accommodate the companion crops. The planting rate depends upon the spacing and the weight of the cuttings or rooted slips used.It is modified in crop mixtures or inter-cropping with other forage crops.

Nutrient management

Farm yard manure @ 25 t ha^{-1}, and $P_2O_5$and K_2O @ 50 kg ha^{-1}, each may be applied at the time of land preparation. Apply N @200 kg ha^{-1} in two or three split doses followed by gentle raking, if possible.

Irrigation practices

The field should be provided with good drainage during the rainy season, as the crop cannot withstand water stagnation.Frequency of irrigation depends upon the rainfall and weather conditions.

Inter-cultivation

Early inter-cultivation once or twice is necessary before the plants establishand grow vigorously. Subsequently, inter-cultivation should be given as and when necessary.

Harvesting

The first cut is taken 9-10 weeks after planting. Subsequent cuts are taken after four to six weeks or when the plant attains a height of 1.5 m. Annually at least six to eight cuts are possible. In order to encourage quicker regeneration from the basal buds, stubbles of 10-15 cm is left out at harvest. Green fodder yield ranges 200-250 t/ ha^{-1} per year from 6-8 cuttings.

Inter-cropping

The grass can be inter-cropped with legumes, such as cowpea, Calapogonium, Centrosema and Glycine legumes. Inter-cropping with legumes improves the quality of fodder.

Nutritional value

Hybrid Napier contains about 10.2 per cent crude protein and 30.5 per cent crude fibre.The leaves are larger and greener, the sheaths are softer and the margins less serrated and hence, the herbage is more palatable. It is juicier and succulent at all stages of growth.It is less fibrous and more acceptable. The oxalate content of some of the varieties maybe high. It can be mitigated, if harvested at longer intervals (45 to 60 days).

Preservation

The grass is ideal for green fodder, silage and hay. The fodder may be cut into pieces using a chaff cutter and ensiled. Urea 2% (Ishrath and Thomas, 2016) and urea 1% (Ishrath *et al.*, 2018) has been found as best additives for ensiling. For organic silage, jaggery 2% can be used (Ishrath, 2016). Legume fodders may be mixed with the grass in the ratio of 1:2 to produce balanced silage. The fodder can also be converted into hay during the dry summer periods. The chaffed material is exposed in the sun by spreading uniformly for a day. Further drying is done under the shade to preserve colour. The quality of the silage or hay remains more or less the same as green fodder.

2. Fodder trees

The leaves, twigs, tender shoots of fodder trees are useful for animals. Fodder trees can be utilized as green fodder, hay or silage. During the dry season, the fodder trees will provide nutritious green fodder while the herbaceous pasture remains dormant and unproductive. The excess green fodder can be used as mulch to conserve the soil moisture as well as to improve the nutrient content also. The important fodder trees are subabul, hedge Lucerne, agathi and shevri.

2.1. Subabul *(Leucaena leucocephala)*

Subabul is a perennial hardy evergreen shrub. It has deep and strong taproot and even the seedlings are deep-rooted. The leaves are bipinnate, 15 to 20 cm long with 10 to 15 pairs of pinnate leaves. Inflorescence is globular and the flowers are white.

Cultivation practices

Climate

Subabul is best suited for warm regions and grows well between 22 and 30°C in regions of 500 to 2000 mm annual rainfall. Because of its strong and deep-root system, the tree is highly drought resistant. It is restricted to elevations below 500 m but it withstands variations in rainfall, sunlight, windstorm, slight frost and drought. It cannot withstand water-logging.

Soil

It requires a deep well-drained neutral soil and can tolerate saline and acid soil. It can also be grown in steep slopes, hilly terrains, gravelly areas and sandy loams.

Season

Planting of seedlings can be done with the onset of rains in May-June or September-October.

Planting material

Seedlings are commonly used for planting. Seed viability is high, but possess dormancy because of hard seed coat. To hasten germination seeds are to be dipped in concentrated sulphuric acid for four minutes and then washed or put in hot water at 80°C for four minutes. Sundry the seeds afterwards for about one hour before sowing.

A seed rate of 3-4 kg ha^{-1} is recommended. Sowing is preferably done during February-March in a nursery or polythene bags or *in situ* at 2-3 cm depth. Give irrigation, if there is no rain. Seedlings (1.5 to 3 months old with 6-8 leaves) are planted in the main field.

Spacing

A spacing of 1 m x 0.1 m is recommended for a pure crop of fodder, 1.5 m x 0.2 m for planting in boundaries and borders of coconut gardens and 2 m x 0.2 m, when raised along boundaries.

Nutrient management

A basal application of N: P_2O_5: K_2O @ 20:50:30 kg ha^{-1} and 5 t/ha farm yard manure is recommended.

Inter-cultivation

Since the early growth of the crop is slow, the tender plants are to be protected from aggressive weeds. Two or three inter-row cultivation is essential to check weeds in early life. Once established, even vigorous grasses seldom smother the plants.

Harvesting

Subabul starts flowering at 125-150 days after planting. First cutting is done after 5-6 months at a height of 70-80 cm from the ground level at a time, when the plants reach a height of 1.5-1.75 m. Subsequent harvests can be made at 50-60 days interval depending on the re-growth. When planted in boundaries, the main shoot is not cut; only side branches are cut for fodder, leaving the top three branches.

In gravelly soil and in low rainfall areas, a yield of 25-30 tonnes per ha per year may be obtained. The irrigated crop may produce 100 t $ha^{-1}$of green fodder per year in seven to eight cuttings.

Nutritional value

Subabul is a highly nutritious leguminous tree fodder with 27-34 per cent protein. The fodder is rich in carotene and vitamin A. Pro vitamin A content is the highest among all plant species. The foliage contains an uncommon amino acid, mimosine, which is toxic to non-ruminants at levels of about 10 per cent of the diet.

Inter-cropping

Leucaena combines well with many grasses like guinea, pangola, dinanath, Hybrid Napier, etc.

3. Fodder legumes

Legumes are important in supplying the most demanding and quality nutrients like protein, minerals and vitamins to the animals. Leguminous forage can be used as supplement with straw-based diets for ruminants in order to improve digestibility of feed and overall performance of ruminants. Inclusion of fodder legumes in the fodder production system is the most efficient way to increase herbage production and, quality and the most economic feed supplement than commercial concentrates. Legumes can be preserved as hay or silage in combination with other grasses. The important fodder legumes suitable for Kerala are fodder cowpea, rice bean and stylo.

3.1. Fodder cowpea (*Vigna unguiculata)*

Cowpea (*Vigna unguiculata*(L.) Walp) is a versatile short duration leguminous plant, commonly grown as grain pulse, vegetable and fodder in semi-arid and humid tropics. It is quick growing and excellent in forage quality. Its value lies in its high protein content and ability to tolerate drought. It also possesses the ability to fix atmospheric nitrogen through its nodules and thereby grows well in poor soils. Cowpea can also be considered as a complimentary crop in the rotation sequence of rice- rice-cowpea because of the leguminous organic residues available for fertility enrichment.Cowpea is a valuable component of farming systems in many areas because of its ability to restore soil fertility for succeeding cereal crops grown in rotation with it (Sanginga *et al.*, 2003). In Kerala, it is raised as rainfed crop during May and also, as a summer crop in rice fallows.

Varieties

Aiswarya

Aiswarya is an improved variety of fodder cowpea developed by hybridization and selection with a green fodder yield of 29.92 t/ha. The crude protein content is 18.5 % and crude fibre content is 20%. It is recommended for cultivation in uplands and homesteads of the three southern districts of Kerala (Abraham *et al*, 2015).

A number of other varieties like Karnataka local, RS-9, UPC-1956, UPC-5287 and UPC-9805 are recommended for cultivation for fodder purpose.

Cultivation practices

Climate

Cowpea is best suited for moderately humid areas of the tropics and subtropics. It usually grows in latitude between 30° N and 30° S and upto 1500 m elevation. The plant cannot withstand frost, excessive and prolonged water-logging, while some varieties are resistant to heavy rains. The optimum temperature required for its growth varies from 15 to 27°C.

Soil

Cowpea can be grown on a wide range of soils from heavy to sandy loam with a pH range of 5.0 to 6.5. Saline, alkaline or water-logged soils should be avoided. Heavy clays encourage vegetative growth with less seed production.

Season

Cowpea for fodder purpose can be grown in any month, if irrigation facilities exist. In Kerala, it is raised as a rainfed crop during May and also as summer crop in rice fallows.

Land preparation

Two to three ploughings are required to produce a coarse seedbed for the crop. Shallow furrows at 3 m apart for leading irrigation water may also be provided. It can be broadcasted or drilled in lines. For seed crop, line sowing is preferred.

Seed rate

Seed rate recommended is 40 to 50 kg ha^{-1} for a broadcasted crop and 15 to 40 kg ha^{-1} for drill sown crop.

Spacing

For drilling, spacing of 30 to 40 cm between rows and 6 to 15 cm between plants is recommended.

Nutrient management

For rainfed crop, at the time of land preparation, FYM @ 10 t ha^{-1} is applied and basal application of N, P_2O_5 and K_2O @ 25, 60 and 30 kg ha^{-1} is recommended. For irrigated crop in addition to the basal dose of 40: 30: 30 kg N: P_2O_5: K_2O ha^{-1}, topdressing of N and K_2O each at 10 kg ha^{-1} after each cut is to be given.

Irrigation practices

A pre-sowing irrigation is important for the proper germination of the crop. If there is lack of soil moisture, shallow irrigation at 3-4 cm depth once in 15 days during summer and once in a month during post-monsoon period is good.

Inter-cultivation

One or two weeding may be required in the early growth stages to combat weed problems. Usually high seed rates are effective in smothering weeds.

Harvesting

As a fodder crop, the first cutting can be given 45 days after planting and subsequent two cuttings at 30 days intervals. A single cut crop yields 25 to 30 t ha^{-1}, whereas green matter yield of 40 t ha^{-1} is obtained from multi-cut cowpea.

Inter-cropping

Cowpea is tolerant to moderate shade. So, it can be successfully grown as inter-crop with maize, sorghum, bajra, guinea grass, Napier grass, etc. to get high yields. In Kerala, cowpea is also raised as an inter-crop in coconut gardens.

Nutritional value

The feeding value of cowpea forage is high. It is superior to other legumes like soyabean because of its low fibre content and minimum wastage in feeding livestock. It has about 16 per cent crude protein and 20 per cent crude fibre.

Preservation

Cowpea is used as fodder crop for green feeding, hay-making, grazing and also for ensiling in mixtures with sorghum or maize. The grains are used as human food as well as animal feed.

3.2. Rice bean

Rice bean [*Vigna umbellata* (Thunb.)] is an important kharif crop of North-Eastern hill region of India. It is used for different purpose, such as food, fodder, animal feed, green manure, cover crop and to maintain the soil fertility. The approximate bio-chemical analysis of fodder at flowering stage reveals 24% dry matter, 15% crude protein, 32% crude fibre, 1% fat, 42% N-free extract, 11%ash, 1.2%calcium and 0.4% phosphorus (Protabase, 2012). It has been identified as the most promising fodder crop during lean periods which is palatable, highly nutritious and is richer than other leguminous fodders like cowpea and moth bean in protein, calcium and phosphorus with more fodder yield (Bhattacharya and Mukherjee, 1998).

The cultivation practices of rice bean are similar to cowpea. The important variety of rice bean released from Kerala Agricultural University is Surabhi.

Surabhi

An improved variety of fodder rice bean developed by mass selection from the accession LRB 64.The variety has recorded a green fodder yield of 33.69 t ha^{-1} in uplands and 38.25 t ha^{-1} in rice fallows. It is highly palatable with crude protein content of 18.9% and crude fibre content of 20.0%. It is superior to cowpea in quality and fodder yield.

4. Cereal fodder crops

Cereal fodder crops are actually food crops grown for human consumption. As grains are the economic parts, the left over portions of plants can be used for fodder purpose. But some of the strains and varieties have the capacity to produce more vegetative parts than grains. These are mainly selected for fodder purpose. They are mainly seasonal crops and these produce more biomass within a short period of time. These can also be included in the homestead system as inter-crops. Important cereal fodder crops are fodder maize and fodder sorghum.

4.1. Fodder maize (*Zea mays*)

Maize or corn is one of the most important cereal crop of the world used both, as food and feed. It ranks third as a food crop after rice and wheat. When grown as a fodder, it gives huge quantities of green herbage in a short time. Fodder maize is usually grown in mixed cropping system or in rotation with several compatible grasses and legumes.

Varieties

Hybrid varieties are Deccan, Ganga-5, Ganga safed-2, and Ganga-3 and composite variety Vijay.

Cultivation practices

Climate

Maize grows best in warm climate, where the day temperature is fairly light. Heavy rains and dry hot winds are not suitable. Favourable annual rainfall is 60-100 cm. The crop comes up well in soils with good drainage and fair moisture status.

Season

The optimum season for sowing is the last week of June to second week of July and September to October. The crop can be raised throughout the year in areas, where irrigation facilities are available.

Land preparation

The land is ploughed two or three times and beds and channels are formed. Seeds can be either broadcasted or dibbled at a spacing of 30 cm between rows and 15 cm between plants.

Seed rate

Seed rate for broadcasting is 80 kg ha^{-1} and for dibbling 40-60 kg ha^{-1} (to be dibbled at 5-6 cm depth @ two seeds per hole).

Nutrient management

FYM @ 10 t ha^{-1} may be applied at the time of preparation of land as basal dressing. N, P_2O_5 and K_2O at the rate of 120, 60 and 40 kg/ ha, respectively, may be given as topdressing. Weeding may be done according to necessity.

Harvesting

First cut of maize can be taken after 60 days of planting or at the milky stage of the crop. A second cut can also be taken, if there is sufficient moisture in the soil.

Success story

Arundev, a dairy farm owner of Thalayil village, Trivandrum has started his Nandini farm 10 years back. Though he is a 50 years old retired army man who returned back after his 16 years of service had great interest in agriculture. After his service, he started cultivating different varieties of banana and vegetables in his 5 acres of land. In 2008, he was much more interested to start a dairy farm. With the help of dairy development department's Milk shed development programme, he has started his dream project with 10 cows. Now he has 70 cows which include HF, Jersy, Brown Swiss, Vechur, Kasaragod Dwarf, etc. Along with dairy farm, he has included fodder crops to his field. Now he has extended the fodder cultivation for 3.5 acres. The fodder cultivation became a successful venture and it helps to feed the cows with good quality green fodder. Then he started selling of planting materials to the farmers and he has one of the Government certified nursery in Trivandrum district. Now he is producing planting materials of different fodder crops and getting an additional profit of one lakh rupees per annum from nursery itself.

References

Abraham, M., Thomas, U.C., and Lakshmi, S., 2015. Improved fodder varieties released from Kerala Agricultural University, Kerala Agricultural University, Thrissur.

Bhattacharya, N. and Mukherjee, A.K., 1998.Herbage growth of rice bean under different dates of sowing and levels of phosphorus.*Forage Research*.24 (2): 121-123

Ishrath, P.K., 2016.Cutting intervals and additives for quality silage production, M.Sc. (Ag) Thesis, Kerala Agricultural University, Thrissur, 105p.

Ishrath, P.K. and Thomas, U.C., 2016. Effect of different additives on quality of hybrid napier silage. *Adv. Life Sci.* 5(12): 5224-5226

Ishrath, P.K., Thomas, U.C., and Ganesh, D., 2018. Impact of additives on quality silage production[abstract], In: *Abstracts,* International Conference on Advances in Agriculture andAllied Science Technologies for Sustainable Development; 10-11, February, 2018, Osmania University, Hyderabad, Telangana, 630p.

Lakshmi, S., George, A., Pillai, G.R., and Rao, N.R., 2007. Nutrient needs of coconut based fodder production systems in homesteads of Kerala. *Better Crops*. 91(4): 28-29.

Posler, G.L., Lensen, A.W., and Fine, G.L., 1993. Forage yield, quality, compatibility and persistence of warm season grass legume mixtures. *Agron. J.* 85: 554-560.

Protabase, 2012.Record display, www.protabase.org

Sanginga, N, Dashiell, K.E. , Diels, J., Vanlauwe, B., Lyasse, O., Carsky R.J., Tarawali, S. Asafo-Adeli, B., Menkir, A., Schulz, S., Singh, B.B., Chikoye, D., Keatinge, D., and Ortiz, R. 2003. Sustainable resource management coupled to resilient germplasm to provide new intensive cereal-grain-legume-livestock systems in the dry savanna. *Agric. Ecosyst. Environ.* 100: 305-314.

Thomas, U.C., Abraham, M., Ishrath, P.K., Fayique, A.C. and Chacko, A., 2018. Coconut based fodder production systems in Kerala. In: Jindal, Y., Chhabra, A.K., and Roy, A.K. (eds.), *Fodder Crops; Approaches for Value Addition and Enhancing Income.*Earth vision publications, Haryana, 233p.

❑❑❑

III. Crops Management

Chapter 19

Nutrient Deficiency Disorders in Plants

Dr. Roy Stephen (Professor) and
Gayathri Rajasekharan (Research Scholar)

Department of Plant Physiology, Kerala Agricultural University, College of Agriculture Vellayani, Thiruvananthapuram-695 522, Kerala

Thirteen mineral nutrients are essential for growth and development of plants. These elements are considered essential regardless of their vast differences in their amounts absorbed by the plants but because of their roles in regulating plant growth and development. Limitation of any one nutrient is adequate enough to decrease growth and productivity. Identifying nutrient deficiency in crop plants and bringing out corrective measures will greatly improve productivity.

Nitrogen

Nitrogen is the most important mineral nutrient required in large quantities. It's a component of amino acids, proteins, nucleotides, chlorophyll, etc. Hence, nitrogen deficiency causes stunting of plant growth and also causes general chlorosis. Nitrogen being a mobile element, the yellowing appears in older leaves. Nitrogen deficient plants look thin, conspicuously pale and will be dwarf.

Phosphorus

Phosphorus is a component of ATP, nucleotides, sugar phosphates, etc. and its deficiency causes dark or dull green leaves alongwith pink colouration due to anthocyanin accumulation. Phosphorus being a mobile element the deficiency appears on the older leaves. Deficiency also causes poor root growth and delayed reproductive development. Sickle leaf disease is characterised by phosphorus deficiency, which results in chlorosis to main veins followed by leaf asymmetry.

Potassium

Potassium is associated with regulation of osmotic potential of plant cells. It's an activator of more than forty enzymes. The deficiency appear as marginal necrosis which starts from leaf tip and margins towards centre of the leaf (inverted 'v' shape) of older leaves since potassium is a mobile element. Rosetting and die back are the two disorders associated with Potassium deficiency.

Calcium

Calcium is involved in the synthesis of middle lamella of cell wall and function as a second messenger in signal transduction pathway of various plant developmental processes. Calcium deficiency causes necrosis of meristamatic region and leaf tip, causing deformed young leaves. Calcium is an immobile element. Deficiency of calcium causes tip hooking, a disease commonly found in cauliflower, beet and tobacco characterised by hooking of leaf tip because of unequal growth in marginal and central region of leaf. White pipe disorder in banana is also caused due to calcium deficiency.

Magnesium

Magnesium is an important component of chlorophyll pigment and also an activator of many enzymes, especially kinases. Its deficiency causes chlorosis between the leaf veins of older leaves since magnesium is a mobile element. Sand-drawn disease is found to be associated with magnesium deficiency which is distinguished by loss of colour at the tips of lower leaves and the veins remain green.

Sulphur

Sulphur is a component of amino acids, co-enzymes and vitamins. The deficiency causes stunting of plant growth and general yellowing as in the case of nitrogen deficiency. Sulphur being an immobile element deficiency appears on the young leaves. Sulphur yellow disease is observed in plants growing in sulphur deficient soils. Tea yellow is caused in tea plants due to sulphur deficiency.

Iron

Iron is involved in the synthesis of chlorophyll molecule and is a component of cytochromes associated with electron transfer chain.

The deficiency of iron causes inter-venal chlorosis of young leaves since iron is an immobile element. Lime induced chlorosis is found in iron deficiency.

Manganese

Manganese is involved in photosynthesis (photolysis) and is an activator of many enzymes. The deficiency of manganese causes mottled chlorosis in the middle leaves. Commonly associated diseases of manganese deficiency are grey spec, pahla blight of sugarcane, marsh spot of pea and speckled yellow of sugar beet.

Zinc

Zinc is involved in the bio-synthesis of plant hormone auxin from the amino acid tryptophan, hence zinc deficiency causes a reduction in the inter-nodal length and cause little leaf symptom. In some plants, the deficiency causes inter-venal chlorosis. *Khaira* of paddy, white bud of maize, rosette of fruit trees and frenching of citrus are the disorders observed in zinc deficiency.

Copper

Copper is associated with enzymes involved in electron transport. The deficiency causes malformed leaves which are twisted or curled upwards. Exanthema or die back of fruit trees and reclamation disease are characterised by copper deficiency.

Boron

Boron is involved in carbohydrate translocation, pollination, pollen germination and fertilization. Boron deficiency causes necrosis of young leaves and terminal buds. It may also cause inflorescence necrosis, crown choking, fruit crack, immature nut fall, etc. in different species.

Molybdenum

Molybdenum is a component of the enzymes nitrogenase and nitrate reductase which are involved in nitrogen metabolism. It is the least abundant nutrient element in plants. Molybdenum deficiency causes chlorosis between the leaf veins and necrosis of younger leaves. Whiptail of brassica and scald of legumes are the deficiency disease observed in molybdenum deficiency.

Nutrient deficiency symptoms in plants

Potassium deficiency

Phosphorus deficiency

Nitrogen deficiency

Calcium deficiency

Magnesium deficiency

Sulphur deficiency

Iron deficiency

Manganese deficiency

Zinc deficiency

Copper deficiency

Boron deficiency

Molybdenum deficiency

General recommendation for Amelioration of Nutritional disorders

Deficiency	Soil application (kg ha^{-1})	Foliar spray (%)
Nitrogen	i. Application of 50 kg urea (or)	Spraying of 1% urea
	ii. Application of 50 kg DAP	
Phosphorus	Application of 25-50 kg DAP kg ha^{-1}	Spraying of 2% DAP
Potassium	Application of Calcium chloride at 60 to 80 kg ha^{-1} or CAN at 50 kg ha^{-1}	Spraying of KCl 1%
Calcium	Application of Calcium chloride at 60 to 80 kg ha-1 or CAN at 50 kg ha^{-1}	Spraying of 0.5% $CaNO_3$ (or) CaC12 (or) Calcium ammonium nitrate
Sulphur	Application of Gypsum at 50 kg ha^{-1}	Spraying of 0.5% gypsum
Magnesium	Application of magnesium sulphate 25-50 kg ha^{-1}	Spraying of 0.5% $MgSO_4$ + 1% urea
Iron	Application of ferrous sulphate 50-100 kg ha^{-1}	Spraying of $FeSO_4$ 1 to 2%
Zinc	Application of zinc sulphate 25 kg	Spraying of 0.5% $ZnSO_4$
Manganese	Application of 25 kg manganese sulphate ha^{-1}	Spraying of 0.5% $MnSO_4$
Copper	Application of 5 to 10 kg copper sulphate ha^{-1}	Spraying of 0.1% $CuSO_4$
Boron	Application of sodium tetraborate or borax at 10 kg ha^{-1}	Spraying of borax at 0.05%

References

Taiz, L. and Zeiger, E., 2010. *Plant Physiology* (5th Ed.). Sinauer Associates Inc. Publishers, Massachusetts, USA, 778p.

Epstein, E. and Bloom, A.J., 2004. *Mineral nutrition of plants: Principles and perspectives.* (2th Ed.). OUP USA, 380p.

Pandey, S.N. and Sinha, B.K., 2001. *Plant Physiology.* (4th Ed.). Vikas Publishing, 704p.

❑❑❑

Chapter 20

Multi-disciplinary Diagnostic Support to Address Field Problems of Farmers in Kerala (*Karshaka Santhwanam*)

Dr. Radhika N.S. (Assistant Professor)

Department of Plant Pathology, Kerala Agricultural University, College of Agriculture Vellayani, Trivandrum-695 522

The MDDT (multi-disciplinary diagnostic team) started in the year 2010-11 made available from the College of Agriculture, Vellayani in the past had undertaken the task of rectification of field problems faced by farmers under the leadership of the P.I. of the project (Dr. Sam T. Kurumthottical) and is continuing. The feedback and review of visits were encouraging and the appreciation from Government and University were encouraging. The project played a pivotal role in solving many field problems, which resulted in the enhancement of productivity in the State and more particularly in homestead farming. The service of the Project and MDDT had been utilized by Government of Kerala in the past. Full technical advice to farmers in the field (Farm Advisory Service) is rendered at all times of need with the support of Multi-disciplinary team. The objectives of the programme are :

- Strengthen the already existing provisions of the Department of Agriculture in sorting out field problems
- Integrate IPM, INM and other technologies to economize the production
- Help University or Government to tackle emergency field problems
- Promote safer food production
- Provide technical support particularly to residents associations in the town areas in promoting kitchen gardens

- Equip the farming community to meet challenges in hi-tech farming

Many farmers and officials of the Department of Agriculture in Southern districts of Kerala are well aware of facilities available at KAU under the *Karshaka Santhwanam* Project particularly at the end of a call with 9496276991, which reach the Principal Investigator of the project. The call is answered; problems are listened to and if needed, the mobile number of the concerned experts are passed on to the farmer for further discussion and clarity of the issues raised. Based on the gravity of the problem, the concerned expert will provide suitable Tele-solutions. If the totality of the problems can't be conceived well over phone by the experts, and the necessity of a combined intervention by Multi-Disciplinary Diagnostic Team (MDDT) is felt, a field visit is planned.

The disease problems that are frequently met by farmers in the field and their management is given below. Banana, coconut, and vegetables occupy a prime position in the livelihood of a Kerala farmer. The cultivation of vegetables by the farmer is according to the preference of the commodity in the market as well as the better returns. Biotic stress is decisive factor in the increasing the production of a crop. For the management, an integrated approach involving all the principles of plant protection would be the best and sustainable.

Integrated management of diseases in banana

Sigatoka Leaf Spots

The leaf spot of banana caused by *Mycosphaerella musicola* is called sigatoka after the Sigatoka valley in Fiji, where the first outbreak occurred.

Symptoms

In case of yellow sigatoka, the major stages of symptom development are yellow streaks, brown streaks and spots. Yellow streaks appear first on older leaves,which later turn brown and enlarge to form spots with a prominent yellow halo. In the case of severe infection, the dead leaf background is greyish white and the entire apical half to two-third of the leaf turns brown and curls downward.

Management

1. Sigatoka control was carried out during earlier periods (1940-1980) using Bordeaux mixture, petroleum oil, emulsified dithiocarbamate or systemic fungicides.Bordeaux mixture was used along with Triton, which reasonably improved the sticking and spreading qualities of the fungicides. Petroleum oil mist in the combination with foliar fungicides like mancozeb afforded protection for longer periods.
2. Heavy oils like Mobil oil A.F., Gas oil, White oil and mineral oil have been found to be effective, when sprayed at 3ml/L. However, symptoms of oil phytotoxicity occur which consists of varying levels of leaf flecking and general bronzing of leaves.
3. Use of more efficient systemic fungicides gives more effective management. The earlier used systemic fungicides were benzimidazole and tridemorph (0.1%). These penetrate through the upper surface of the youngest unfurled leaves, where the pathogen is in in the early stages of infection. Presently, the triazoles are being effectively used for the management of sigatoka disease in banana and the use of Tilt (Propiconazole 0.1%) has become very popular for the management of these leaf infections. However, the continuous uses of systemic fungicide can result in the formation of the pathogen. This can be avoided by alternating the use of systemic fungicide with 1% Bordeaux mixture.

Fruit Diseases

Cigar End Rot

Pathogen

Verticillium theobromae

Symptoms

When young developing fruits get injured, they take up infection. Infected fruit develop necrosis on the pistillate end and the fruit skin becomes folded and shrunken as the infection spreads. The dead floral parts persist and the powdery conidia form on the shrivelled stalk end of the fruit. The druit pulp undergoes characteristic dry rot and gets converted into dry fibrous mass.

Management

1. Removal of distal bud is recommended after the opening of all the hands. The extent of disease in stored fruits is related to the degree and intensity of latent infection and the storage conditions.
2. Protective sprayings are usually recommended with systemic fungicides, when the fruits is still very young to prevent the development if latent infection.

Viral Diseases of Banana

Bunchy Top

Symptoms

The affected leaf shows green streaks along the secondary vein on the underside of the lamina and on the midrib and petiole. These streaks are about 0.75 mm wide and vary in length upto 2.5 cm. The streaks vary from series of dark green dots to a continuous dark green line often referred to as morse code symptom. The subsequent leaves show the same symptom and get reduced in size progressively showing marginal chlorosis and curling. Plants infected at an early stage of development rarely produce a bunch, while those infected late form distorted bunches.

Epidemiology

All *Musa* species and cultivars tested so far to be susceptible to BBTV. Alternative hosts include *Alpinia purpurata, Colocasia esculenta* (taro), *Canna indica* and *Hedychium coronarium.*

Kokkan disease (*Banana bract mosaic virus*)

Kokkan disease was found to affect other varieties like Palayankodan, Kodappanilla-kunnan, Monthan, Kanchikela, Poovan (Rasthali), Karpooravally, Chenkadali and Nendran. Nendran is a highly susceptible variety.

During the young stage of Nendran banana plant (two months old), pinkish streaks can be seen on the pseudostem. All the kokkan affected plants may not show this symptom, but once this symptom is expressed, there is no doubt that the particular plant is affected. Necrotic streaks are another important symptom of the disease. The streaks are initially brown, which later turn black. It occurs on all aerial parts of the affected plant except on lamina, the length being a few mm to 10 cm. All the kokkan-affected plants will exhibit the necrotic streaks from third month onwards at one stage or the other. Some of the affected plants retain the necrotic streaks throughout the growth period. In certain cases, it disappears with the senescence of the affected portion.

The affected plant produces only a small bunch. The fingers are small, curved and widely divergent with pale green to ashy green colour as compared to healthy ones. The abnormal colour and reduction in the size of bunch depend upon the severity of the disease.

Suckers should not be taken from affected plants, which show necrotic streaks or abnormal colour of the pseudostem. When the young plants show the symptom of pinkish streaks, they should be uprooted and destroyed.

Infectious chlorosis (*Cucumber mosaic virus*)

The disease is noticed in varieties, such as Nendran, Palayankodan, Karpooravally, Kosthabontha, Peykunnan, Bhimkhel, Mottapoovan, Dakshinsagar, Madhuraga, Rasthali and *Musa ornata*.

The most characteristic symptoms are the loss of leaf colour in patches, appearance of parallel chlorotic streaks on the younger leaves giving a stripped appearance of the leaves. As the disease progresses, leaves emerge distorted, margins become irregularly wavy, often with blotches of necrotic tissues and the lamina is reduced in width. In severe cases, rotten areas are found throughout the leaf sheath and pseudostem. The affected plants produce only small bunches. This is a virus disease transmitted by aphids.

1. Use disease free suckers for planting.
2. Eradicate disease affected plants.
3. Avoid growing leguminous and cucurbitaceous vegetables as inter-crop in banana in disease prone areas.

Banana Streak

Pathogen

Banana Streak Virus (BSV)

Symptoms

Chlorotic and necrotic streaking is commonly seen on the affected plants but the severity of the symptoms expression is found to vary with the virus concentration in the leaves. The leaf bases of the affected plants may fall off from the pseudostem. The leaves may progressively turn narrower and thicker. There is general leaf distortion along with improper in furling or choking of newly emerging leaves. Broad yellow lines may appear on the leaf blade parallel to the midrib.

Integrated management of diseases in coconut

Leaf Rot

- Affect palms of all ages, especially in palms below the age of 25 years
- Water soaked lesions on the emerging spindle and on young tender leaves with different shape and colours
- Dark brown spots develop on the tender leaves, which later enlarge and dry up while the leaf emerges
- Successive central shoots are affected, resulting in rotting of all the leaves on the crown
- These lesions enlarge, coalesce leading to extensive rotting of spindle leaves
- Extension of rotting into the interior of spindle resulting its decay and sometimes to rotting of buds.
- Tips of leaflets and midribs often become blackish, shriveled and fall off in wind, giving a 'fan' like or 'arrow leaf' appearance to the leaves.
- Successive infection of the emerging spindles results in appearance of symptoms in most or even in all leaves of the crown.
- Severe lesions and their coalescing on petiole, midrib, midveins of leaflets leads to severe blighting of lamina.
- Breaking of digital ends of leaves, become yellow and eventually hangs, dry and ultimately fall off.

Causal organism: Fungal complex initiated predominantly by *Colletotrichum gloeosporioides*, *Exserohilum rostratum* and *Fusarium* spp.

Management

- Removal and destruction of disease advanced/senile/ uneconomic palms, followed by replanting with healthy seedlings from elite palms and varieties of hybrid tolerant to the leaf rot - root (wilt) disease complex.
- CGD and hybrids of CGD X WCT are tolerance under field conditions.
- Adopt all recommended packages viz. agronomic, cultural, plant protection, etc. as aaplicable to root (wilt) disease endemic region.
- General cleaning of the crown in pre-monsoon period.
- Remove and destroy by burning the rotten portions from the spear and the 2-3 adjascent leaves.
- Pour 300 ml -1000 ml of fungicidal solution of 0.2-0.4% contaf or 0.5 to 1% mancozeb onto spindle region.
- Spray crowns and leaves with 1% Bordeaux mixture or 0.3 to 0.55 Copper oxy chloride formulation or 0.4% manxozeb in January, April-May and September. While spraying, care has to be taken to spray the spindle leaf.
- Basal Stem Rot/Tanjore Wilt/Foot Rot/Ganoderma Wilt/Anabe
- Middle aged palms are seen fatally affected.
- Profuse bearing of the palms just prior to and at the initiation of symptoms.
- Withering, drooping, yellowing and browning of outer whorls of leaves and usually remaining for a long time suspended vertically round the stem to form a skirt of dead leaves around the trunk.
- Breaking or buckling of leaves very close to their bases.
- Successively produced leaves are smaller so that the crown becomes reduced to a mere handful of short erect yellow leaves.
- Tapering of the apex of the trunk.
- Falling off of drooped leaves one by one leaving only a few smaller leaves at the apex.
- The newly formed leaves are pale and smaller in size.

- Finally withering of these smaller leaves and decaying of bud.
- Crown is easily blown off by wind.
- Rotting of the basal portion of the stem is very characteristic.
- Brittle bark often gets peeled off in flakes leaving open cracks and crevices.
- Bleeding of reddish brown viscous fluid with the decayed tissues beneath restricted to basal portion of the stem upto 10-15m height from the base.
- The internal tissues are discoloured and disintegrated, emitting a bad smell.
- Rapid decaying of bole resulting in the formation of large cavities.
- Extensive damage of the root system following rotting and disintegration of cortical tissues.
- Bracket shaped shiny waxy brown frutification of the fungus at the base of the trunk after the death of the palm.
- Necrosis of male flowers and poorly developed few female flowers in the inflorescence, arresting the development of bunches and barren nuts.

Causal organism: Ganoderma lucidum and Ganoderma applanatum in India and G. boninense in Sri Lanka.

Fungus is soil borne inhabiting in dead as well as living woody plants in the soil. Prolonged drought, high soil temperature, low rainfall, extensive flooding, sandy or sandy loam soils, presence of hard pan in the sub-soil, neglected gardens are favourable conditions for the disease incidence.

Disease management

- Complete removal of dead palms and palm in advanced stage of the disease and burning of the boles and root bits of the palms.
- Isolate the affected palm from the healthy ones by digging a trench of 1m deep, 50 cm wide and 1.5 m away from the bole of the infected palm.
- Addition of 50 kg of farmyard manure or green leaves per palm per year.

- Apply 5 kg neem cake per palm per year
- Reduce fertilizer application to one-fourth of the recommended dose
- Drench the base with 40 litres of 1% Bordeaux mixture or 0.1% Calixin after soaking soil at quarterly interval for one year
- Root feeding of 2g Aureofungin -soil + 1g of Copper sulphate in 100 ml of water or Calixin 2 ml in 100 ml water at quarterly interval
- Avoid flood irrigation or ploughing in infected gardens to prevent spread of inoculums.
- Regular basin irrigation during summer months or moisture conservation by coconut husk burial
- Avoid growing leguminous crops in and around the garden.
- Application of *Trichoderma harzianum* fertified in neem cake/ compost/vermicompost/other organic wastes.

Stem Bleeding

- Very common disease in Northern Kerala.
- Exudation of reddish brown liquid through the growth cracks on the basal part of the trunk, which later turn black on drying forming encrustation with brownish orange margin
- Bleeding patches progress both, upwards and downwards and covering major portion/part of the trunk
- Coalescing of lesions/patches to form larger patches
- Decaying of tissues beneath the bleeding points and become yellowish
- Reduction of leaf size leading to reduction in crown size
- The rate of leaf production shows down
- Reduction in bunches production
- Heavy shedding of nuts and buttons

In advanced stages, uniform yellowish and of drooping leaves leading to complete death of the palm:

- Causal organism
- *Theilanopsis paradoxa - Conidial stage (imperfect)*
- *Ceratocystis paradoxa - Ascigerous stage (perfect)*
- The fungus is airborne:
  - Growth cracks on the trunk, sudden heavy manuring or heavy rains followed by extensive dry period, poor drainage, soil moisture stress, hard pan formation in the soil, imbalance in nutrition, excessive soil salinity, stem injury, lightning attack, insect attack, etc. can act as pre-disposing / aggravating factors of this disease.

Management

- Chisel out the affected tissues completely and paint the wound with tridemorph (calixin) 5%.
- Apply coal tar or rubber kote after 1-2 days
- Burn the chiseled materials
- Avoid any mechanical injury to trunk
- Application of neem cake 5kg per palm in the basin along with other manures and fertilizers
- Root feed with tridemorph (calixin) 5% through 3-4 roots during September - October, January - February and April - May
- Drenching the wet basin with tridemorph (calixin) 25 ml in 25 litre of water once in 3-4 months
- Trailing pepper on coconut to prevent direct exposure of cracks to pathogen
- Regular basin irrigation during summer months and moisture conservation by coconut husk burial
- Apply normal dose of organic manures (50 kg/palm/year) and balanced fertilizers.

Sooty Mould

- A black layer of fungal growth covering on leaves and reducing their photosynthetic efficiency.

Causal organism: *Capnodium* sp.

- Fungus develop in the sweet exudation of scales and mealy bugs attacking the tender leaves.

Management

- Spray 1% starch water on the affected leaves
- On drying fungus along with starch coating will be removed as flakes
- Control white flies, scales and mealy bugs using insecticides mentioned earlier.

Diseases of Vegetables

Diseases of Cowpea

Collar rot and web blight

Symptoms

This disease is noticed at all stages of crop growth and causes seed rot, seedling blight, collar rot, root rot and web blight among which, the collar rot and web blight symptoms are more prevalent. Reddish brown water soaked lesions is seen at the base of the stem in contact with soil. Under favourable condition, the lesions extend and cause girdling of the stem. The stem at the collar region becomes sunken and rotten. Thick mycelium and brown scerotia are noticed in the affected portion at later stages. The foliage above the infected stem shows yellowing. Defoliation sets in, after lesion formation.

Web blight is noticed on the foliage. Initially, the symptoms are seen as small necrotic lesions, with brown centers and pale green margins. The leaves show water soaking and rotting. On the rotten areas, white cobweb by mycelial growth is noticed and the adjacent leaves are seen webbed together by the mycelium. Sclerotia are also noticed in the affected areas.

Pathogen

Rhizoctonia solani

Management

1. Cultural practices like shallow sowing, use of raised beds, deep ploughing, mulching and rotation with cereal crops help in decreasing disease incidence and severity. When cowpea is rotated with rice, the severity of disease may be augmented.
2. Application of organic amendments like neem cake.
3. Collection of seed from healthy plants.
4. Roguing of severely infected plants and drenching with copper oxychloride or mancozeb @ 2-4 g/L or carbendazim 1g/L @ 2-2.5 L or fluzilazole 1ml/L or hexaconazole 2ml/L of fungicide solution per plant
5. Seed treatment with carbendazim 1g/kg seed.
6. Soil drenching can be done with biocontrol agent Psuedomonas 20g/l.

Wilt

Symptoms

The pathogen affects the crop at any stage of crop growth. Three different types of symptoms are commonly noticed in cowpea grown in Kerala :

1. Wilt: The plants show yellowing, withering and drooping of leaves. The vines show blackening and drying. The roots show rotting and blackening.
2. Basal swelling: The base of the stem swells up and becomes disintegrated and shredded. The taproot and lateral roots are affected.
3. The stem, leaf petioles, etc. become flattened. The flowers show fasciation and sterility.

Pathogen

Fusarium spp

Management

Same as for collar rot

Leafspot

Symptoms

Reddish brown spots appear on the lower leaves, which spread upwardsresulting in defoliation and drying of vines.

Pathogen

Cercospora sp.

Management

1. The infected leaves and crop debris should be removed and burnt.
2. Spraying with copper oxychloride or mancozeb @ 2-4 g/L or carbendazim 1g/L @ 2-2.5 L or fluzilazole 1ml/L or hexaconazole 2ml/L of fungicide solution per plant
3. Seed treatment with carbendazim 1g/kg seed.
4. Spraying biocontrol agent Psuedomonas 20g/l.

Rust

Symptoms

Rust occurs in the areas of high humidity and low temperature. The leaves, leaf petioles and stem show minute reddish brown raised pustules. Later, appear as dark brown or black. Leaves turn yellow and result in defoliation.

Pathogen

Uromyces appendiculatus.

Management

5. Spraying tridemorph (0.1%) at 45, 60, and 75 days after sowing help to control the disease.
6. The infected leaves and crop debris should be removed and burnt.

Cowpea aphid-borne mosaic

Symptoms

Initially, the young trifoliate leaves show vein clearing. Gradually, chlorotic areas increase to cover the entire leaf. Soon leaves show mosaic mottling, vein banding, puckering and distortion. Severe infection results in considerable reduction of the size of new leaflets and also blistering and bleaching of leaves. The pods are also twisted, curved and reduced in size. The seeds in these pods are shrivelled and few in number.

Pathogen

Cowpea aphid-borne mosaic poty virus/ Black eye cow pea mosaic poty virus. The virus is transmitted by aphids.

Management

1. Collect seeds from healthy, disease free plants.
2. Manage vectors by spraying insecticides like malathion 50 E.C. (2ml/L), quinalphos 20 E.C. (1.25 ml/L), or, neem oil emulsion (This is done by dissolving 60 g bar soap in 500 ml water and blending 1 L. neem oil into it and diluting the mixture 10 times). Spraying should be directed to the under surface of leaves and other parts of the plant.

Cowpea bud necrosis

Symptoms

Characteristic symptoms are development of typical chlorotic spots on the leaves, as the disease advances, severe necrosis of the leaves is noticed. Veins and veinlets also show necrotic symptoms. Necrotic symptoms are also noticed in stem, growing point, buds, pods, etc. Finally, the entire plant dries up.

Pathogen

Groundnut bud necrosis virus

This virus is transmitted by different species of thrips, such as *Scirtothrips dorsalis* and *Thrips tabaci* and by mechanical means.

Management

Same as Cowpea aphid-borne mosaic

Diseases of Amaranthus

Amaranthus leaf blight

Symptoms

The disease starts off as small straw coloured lesions in the foliage of young plants, which later coalese and turn greenish and often get dislodged from the leaf lamina causing shot holes and disfiguring it entirely. The infected foliage cannot be marketed and thus, reduces the commercial value of the crop considerably.

Pathogen

Rhizoctonia solani

Colletotrichum gloeosporioides

Management

1. The green type, CO–1 is resistant to the pathogen and it can be used as crop mix to reduce the spread of the pathogen under field conditions.
2. Prophylactic spray application of baking soda with turmeric powder at a ratio of 1: 4 can give rise to systemic induced resistance in the plants.
3. Fungicidal control can be effectively done by the periodic application of mancozeb@ 4g/L using cowdung supernatant as the diluent.
4. Spraying with Psuedomonas 20g/L

Cowdung supernatant can be prepared by dispensing 1kg of fresh cowdung in 10L of water and taking the supernatant after sieving through fine muslin cloth. The fungicidal solution has to be prepared in the cowdung supernatant and sprayed as fine droplets on to the under side of the foliage. This spraying has to be ceased 14 days before harvesting of the crop to minimize residue problem.

White Rust

Symptoms

Creamy white blisters or raised pustules are noticed on the under surface of the leaves, which cause formation of pale raised areas on the upper surface of leaves. When the infestation is severe, the leaves dry up and defoliate. The disease is more severe during the cooler months.

Pathogen

Albugo bliti

Management

The fungus is disseminated under field condition by wind splashed rains or irrigation. It can be controlled by the spray application of 1% Bordeaux mixture or 0.4% copper oxychloride.

Diseases of Bhindi

Yellow vein mosaic

Symptoms

Vein clearing and inter-veinal chlorosis are the typical symptoms of the disease. The yellow network of veins becomes very conspicuous and the veins and veinlets are thickened. The fruits are reduced in size, malformed and stunted.

Pathogen

Yellow Vein Mosaic Virus. The virus is transmitted in nature by white fly (*Bemisia tabaci*).

Management

Mulching of the crop beds with lemon grass refuse serves as an effective repellent.

Varieties like Punjab Padmini, Arka Anamika, Susthira and Varsha Upahar are tolerant to the disease.

Cercospora Leaf Spot

The fungus grows on the under surface of the leaves in an angular manner as a sooty, dark olivaceous growth. When there is severe infection, defoliation results.

Pathogen

Cercospora abelmoschi and C. *malayensis*.

Management

Periodic application of copper oxychloride 0.2% or mancozeb 0.3% can manage this problem effectively.

Disease of Cucurbitaceous Vegetables

Fruit Rot/ Cottony Leak

This is a very common disease affecting cucurbitaceous vegetables including bitter gourd, snake gourd, pumpkin, bottle gourd, ash gourd etc. during hot summer months and during rainy season. Once the fruits are infected in the field, they continue to rot during transit and storage too.

On the fruits, infection occurs around wound sites due to insect bites, soil particles and excessive wetness. The diseased fruits appear to be covered with luxuriant cottony growth on its surface and it starts to soften considerably. The disease starts as soft dark green water soaked lesions that gradually develop to form watery soft rot. In watermelons, the decay usually starts at the blossom end. The infected portion decomposes rapidly.

Pathogen

P. aphanidermatum, P. butleri and *Sclerotium rolfsii*

Management

1. Crop rotation.
2. Do not allow the fruits to come in touch with the soil from which the pathogens gain entry.
3. Mulch with straw/paper to reduce infection.

Downy Mildew

The disease occurs in a very serious form, when there are intermittent rains during the cropping period.

Symptoms

The fungus cause pale yellowish, angular patches on the leaves, which become dark brown and dry up later. White downy growth of the fungus is visible on the lower surface of the affected leaves during humid conditions. During the later stages, the entire leaves are affected and may turn yellow, decay and fall off.

Pathogen

Pseudoperonospora cubensis.

Management

1. The crop has to be protected from the pathogen for which weekly sprays of mancozeb 0.3% can be given starting at the appearance of the disease. The interval between the sprayings can be adjusted based on the disease severity.
2. Remove and destroy the severely infected leaves/vines before application of the fungicide.

Powdery Mildew

This is a very serious problem in cucurbits during the hot season.

Symptoms

The foliar symptoms are initiated as whitish grey superficial spots on the upper surface. These spots enlarge forming large patches of powdery growth of the fungus. When the attack becomes severe, defoliation occurs. Owing to infection, considerable yield reduction results.

Pathogen

Erysiphe cichoracearum

Sphaerotheca fuligena

Management

1. Morestan or Karathane 0.2% spray can be used to manage the disease, when it appears in a severe form.
2. Spray with carbendazim 0.1% or thiophanate methyl 0.1%.
3. All the affected plant parts have to be removed and destroyed to maintain crop hygiene.

Anthracnose

Symptoms

All aerial parts of the cucurbit plants are affected. On the foliage, symptoms appear as yellowish water soaked region that later turn brown or black and dries up. When fruit pedicels are infected, the young fruits may darken, shrivel and die. Rounded, dark sunken spots develop on the affected tissues. When moisture is present, the fungus starts sporulating along the lesions to produce masses of salmon pink coloured spores. Cankers lined by the spore masses are commonly seen, wherever the disease is prevalent.

Pathogen

Colletotrichum lagenarium

Management

1. Spraying of the crop, starting at the two leaf stage at ten days intervals with 0.2% mancozeb or 0.1% benomyl or carbendazim, is effective in reducing field infection.
2. Dry seed treatment with carbendazim at 2.5g/Kg seed is useful for the management of seed borne infection.
3. The diseased plant refuse and fruit rind should be destroyed to maintain crop hygiene.

Fusarium wilt

Wilt of cucurbits is very common in the tropics and there are different forms of the fungus *Fusarium oxysporum* causing the disease in different cucurbits.

Symptoms

The fungus attacks the plants during all stages of growth. Pre-emergence damping off and post-emergence seedling mortality also can occur. In the older plants, the characteristic wilt symptoms appear when the leaves lose their turgor and hang down during the hot periods. The wilting progresses gradually and cause characteristic tip burn symptoms on the foliage of the affected plants. During wet conditions, dead stems are covered with pinkish white masses of the pathogen.

Management

The seed borne inoculum can be managed by dry seed treatment with carbendazim at 2.5g/Kg seed. Liming at the rate of 100g/pit, followed by regular application of small quantities of well-decomposed neem cake per pit is helpful in the management of the problem.

Mosaic

Mosaic diseases affecting cucurbits are very much prevalent in India.

Symptoms

If early infection is present, the symptoms are very severe with total yield loss. The typical mosaic mottling appears on the young upper leaves, which gradually cover the entire foliage. The diseased leaves become mottled, deformed and reduced in size and are curled unnaturally often turning brittle. The internodes are shortened leading to a bushy nature of the infected plants. Young fruits are rough, mottled and deformed. Systemically, infected vines do not normally give any significant yields.

Pathogen

Cucumber viruses 1, 2 and 3. Most of the common aphids transmit the virus including *Aphis gossypii, A. craccivora, Myzus persicae,* etc. This virus is also sap transmissible.

Yellow vein mosaic

This has been reported in South India, which has become very prevalent on pumpkin and other cucurbits in this region.

Symptoms

The virus causes severe yellowing and stunting of plant leading to total yield losses as in the case of the yellow vein mosaic of bhindi.

Pathogen

The Yellow vein mosaic virus. The virus is transmitted by the white fly, *Bemisia tabaci.*

Management

1. The virus infected plants should be rogued and destroyed.
2. Periodical application of neem seed emulsion 2% can keep the insect vectors from causing the rampant spread of the disease.

Phyllody

Phyllody disease is noticed in bitter gourd, snake gourd and cucumber. The disease is characterised by shortened internodes and phylloid flowers. All the floral parts, such as calyx, corolla, male and female reproductive structures are transformed into green and leaf like structures. Such plants do not bear fruits.

Pathogen

Phytoplasma. Disease is transmitted through grafting and leaf hopper vector.

Management

1. Remove all the infected plants.
2. Periodical application of neem seed emulsion 2%.

Diseases of Solanaceous Vegetables - Tomato, Brinjal and Chilli

Damping off

This is a problem that occurs in most of the vegetable crops, especially when they are raised in a nursery. It is responsible for the poor stand of crop in the nursery beds.

Pre-emergence damping off occurs, when the seedlings are killed before they emerge from the soil. Since the entire process occurs below the soil level, result in the low germination rate of the seed.

When the seedling stand is very thick and when there is overhead irrigation with chances of water stagnation, post-emergence damping off results. The infected tissue becomes soft and water soaked and seedlings fall over in groups. As the disease spreads rapidly, large patches of diseased areas develop soon.

Pathogens

Pythium spp. viz; *P. aphanidermatum, P. debaryanum, P. butleri, P. ultimum* and *P. arrhenomanes.*

Septoria leaf spot of Tomato

This is an important disease causing very heavy defoliation in its severe form. The initial symptoms start off as small spherical spots on the under surface of the older leaves, which later enlarge to develop white or grey centres with dark brown margins. In susceptible cultivars, the spots enlarge rapidly and entire leaflet die and fall off. During continuous rainy weather, total defoliation may occur.

Management

1. Follow crop rotations avoiding other solanaceous crops.
2. Remove and destroy the infected crop debris.
3. Periodically spray mancozeb 0.3% to alleviate the disease.

Powdery mildew

Symptoms

Whitish powdery patches appear on the lower surface of the leaf, stem and fruits. The corresponding upper leaf surface show bright yellow spots. There is severe defoliation.

Pathogen

Leveillula taurica.

Management

1. Proper field sanitation has to be done to remove collateral weed hosts.
2. Spraying with wettable sulphur 0.2% or benomyl/ carbendazim @ 1g/L is effective.

Anthracnose and fruit rot of Chilli

Symptoms

Anthracnose causes necrosis and die back of twigs from tip downwards. The fruits are infected, when they turn ripe and red. The disease begins as a small dark spherical spot on the skin of the fruit, which spreads along the length of the fruit. On the discoloured areas, numerous acervuli of the fungus are seen. When a diseased fruit is cut open, the lower surface is covered with black stromatic tissues or sclerotia of the fungus. In advanced stages, the seeds are covered by a mat of fungal hyphae. The fruit infection that starts in the field will continue even after they are dried and stored leading to massive losses during storage.

Pathogen

Colletotrichum capsici.

Management

1. Dry seed treatment with carbendazim 2.5g/kg seed can effectively manage the seed borne infection.
2. The anthracnose can be properly managed by the periodic application of copper oxychloride 0.3% or mancozeb 0.2% or carbendazim 0.1% in a tandem manner one after another at bi-weekly intervals.
3. Maintaining field sanitation and crop hygiene is very important.

Phomopsis blight and fruit rot of Brinjal

This starts off as foliar blight, the most destructive phase is fruit rot. The fruit rot phase though incited in the field itself often causes heavy losses in transit and storage.

Foliar infection is manifested as circular grey to brown spots on the leaves nearer to the soil level. Lesions are formed on the stem, which turns grey in the centre. The fruits are attacked which may eventually cover the entire fruit. If the entry point of the pathogen is through the calyx, the whole fruit becomes mummified due to dry rot.

Pathogen

Phomopsis vexans.

Management

1. Seed treatment with carbendazim 2.5g/kg seed helps in reducing the seed borne inoculum.
2. Once the disease appears in the field, it can be kept under check by giving periodic spray with captan 0.2%, carbendazim 0.1% or mancozeb 0.3%.

3. The varieties Pusa Bhairav, Pant Samrat and Haritha are tolerant to the disease.

Bacterial Wilt of Tomato, Chilli and Brinjal

Symptoms

The symptoms include wilting, stunting, yellowing of the foliage and collapse of the whole plants. Often there is downward curling of the leaflets and excessive production of adventitious roots along the stem. The entire vascular system undergoes blackening and often bacterial slime oozes out from it, if cut open.

Pathogen

Ralstonia solanacearum.

Management

1. The incorporation of bleaching powder into the soil at the rate of 15 kg/ha was effective in reducing disease level and increasing yield.
2. The application of well-decomposed neem cake at the rate of 25g per pit in a phased manner can help in improving the soil hygiene and keeping the pathogen level to the minimum.
3. The pits can be prophylactically drenched using 0.3% copper oxy chloride two weeks before transplanting.
4. The resistant varieties may be used for cultivation to keep the disease in check.

   Brinjal - Surya, Swetha, Haritha, Neelima, Arka Nidhi, Pant Samrat

   Tomato - Sakthi, Mukthi, Anagha

   Chilli - Ujwala, Anugraha, Pant C-1, KAU Cluster, White Kandari

Tomato spotted wilt

Symptoms

The initial symptoms include the slight thickening of the veins of the younger leaves. Simultaneously, concentric rings appear in the foliage and the younger leaves develop a tendency to curl downwards and inwards. Later the entire leaves may undergo bronzing or bronzing may occur in patches. During later stages of disease development, distinct mosaic mottling develops on the foliage. The infected fruits develop paleness in colouring, when ripe with typical irregular mottling and distinct concentric circles.

Pathogen

Tomato spotted wilt virus. The common thrip vectors include *Thrips tabaci* and *Frankliniella schultzei*.

Management

1. The infected plants should be systematically rogued and destroyed.
2. Proper management of the insect vectors should be effected by giving periodical sprays using systemic insecticides like dimethoate 0.05%.

Leaf curl of tomato

The characteristic symptom is the curling, puckering and blistering of leaves. The venation on the under surface of the leaves produces outgrowth called enation. The infected foliage exhibits mottling, vein clearing and the plant exhibits complete or partial sterility.

Pathogen

Tomato leaf curl virus. The virus is transmitted by white flies (*Bemisia tabaci*) and through grafting.

Management

1. Apart from the control measures included for the tomato spotted wilt, placement of yellow sticky traps will help in reducing the vector population in the field.
2. Pusa Jwala and Jwalamukhi are resistant to this disease.

□□□

Chapter 21

Micro Irrigation

Dr. Xavier K. Jacob (Professor & Head)

Department of Agricultural Engineering, Kerala Agricultural University, College of Agriculture Vellayani, Thiruvananthapuram-695 522

Drip irrigation also known as *trickle irrigation* or *microirrigation* is an irrigation method that applies water slowly to the roots of plants, by depositing the water either on the soil surface or directly to the root zone, through a network of valves, pipes, tubing, and emitters. Drip irrigation is a type of micro-irrigation system that has the potential to save water and nutrients by allowing water to drip slowly to the roots of plants, either from above the soil surface or buried below the surface. The goal is to place water directly into the root zone and minimize evaporation. Drip irrigation systems distribute water through a network of valves, pipes, tubing, and emitters. Depending on how well-designed, installed, maintained, and operated it is, a drip irrigation system can be more efficient than other types of irrigation systems, such as surface irrigation or sprinkler irrigation.

Micro-irrigation, also called **localized irrigation, low volume irrigation, low-flow irrigation**, or **trickle irrigation** is an irrigation method with lower pressure and flow than a traditional sprinkler system. Low volume irrigation is used in agriculture for row crops, orchards, and vineyards. It is also used in horticulture in wholesale nurseries, in landscaping for civic, commercial, and private landscapes and gardens, and in the science and practice of restoration ecology and environmental remediation.

There are several types of micro-irrigation systems. Many of the components are the same for all of these types of systems. Most systems typically include filters, pipes, valves, and tubing. The main difference is in the type of emission device that is used to deliver the water to the plants. Drip irrigation utilizes drip emitters that deliver water at very low rates. The typical range is 2 to 8 litres per hour. In some systems, the emitters are installed manually on the outside of the tubing and placed, where needed. Other systems might use integral dripper line or drip tape with the emitters already installed at a pre-determined spacing. Micro-sprinklers, which can include fixed stream sprays and rotating spinners, typically deliver water at a higher rate, such as 50 to 100 litres per hour and will cover a larger area than drip emitters. These are more typically used in tree orchards, where the plants are larger. The goal is to distribute water slowly in small volumes and target it to plants' root zones with less runoff or overspray than landscape and garden conventional spray and rotary sprinklers. The low volume allows the water to penetrate and be absorbed into slow-percolation soils, such as clay, minimizing water runoff.

There are a wide variety of system components included in a micro-irrigation systems. Most systems include a filter. These may include pre-filters, sand separators, media filters, screen filters, and disc filters. The level of filtration required depends on the size of the emission device and the quality of the water source. A pressure regulator or regulating valve may be required to reduce the system pressure to the desired level. Automatic or manually operated valves will be required to switch from one irrigated section to another. An irrigation controller will be used with automatic systems and may also be needed for backflushing the filter or sand separator. Since water conservation is a frequent reason for choosing micro-irrigation systems, soil moisture sensors, rain shutoff sensors, and sometimes, even weather stations may be installed.

Emission devices

Microtubing

Microtubing is one of the oldest types of drip irrigation devices and was used in greenhouses in the 1970s. It consists of a very small diameter tubing. Flow is regulated purely by the length and diameter of the tubing. Weights or stakes are sometimes attached to the end of the tubing to keep it in place.

Fixed flow drip emitters

Low-flow irrigation systems in gardens using drip apply water through two methods:

- Pre-installed small holes in small diameter tubes placed on or below the surface, or

- Self-cleaning emitters, in different precipitation rates, pre-installed or contractor installed for different rate emitters on same supply line (i.e. trees-higher, perennials-lower). The Flexible supply pipe can be buried either underground or pinned on the surface and buried under.

Low volume irrigation systems often use the two delivery components of drip systems to apply water through small holes in small diameter tubes placed on or below the surface of the field. This is done instead of agricultural surface irrigation and furrow irrigation for vegetables, fruits and berries, and other high-value crops.

Adjustable drip emitters

Trickle emitters, also called 'spider sprays,' come in fixed or adjustable radius shapes and diameters, and are installed directly on the flexible supply pipe or on tubing connected to it, and mounted on small stakes. Trickle emitter-'Spider sprays' work well for plants with more fibrous root systems, tree and large shrub basins, and in pots and container gardens-allowing automated watering of plants on decks and patios. Mist emitters can also be used in pot, both on the ground and hanging, with humidity-fog watering for epiphytes and ferns replicating habitats.

Micro-sprinklers

Low volume micro-sprinklers may be attached to hard plastic risers or attached to standard sprinkler heads, but are more typically mounted on stakes and attached to small diameter micro-tubing connected to polyethylene tubing with a barbed connector. Some micro-sprinklers have a fixed spray or stream pattern, while others rotate. These are installed above ground and are often used for fruit and nut orchards and vineyards. These systems are expensive, even for large-scale agricultural use, and are pre-dominantly used for high-value crops.

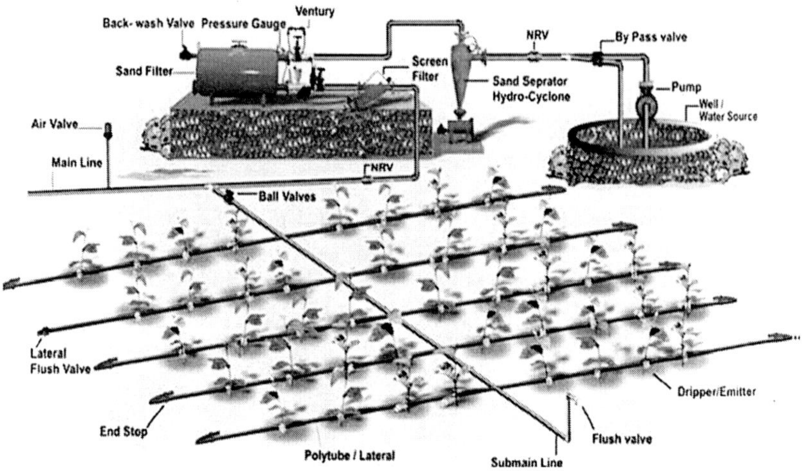

Drip Irrigation Layout

Water Application Characteristics

- Low rates
- Over long periods of time
- At frequent intervals
- Near or directly into the root zone
- At low pressure
- Usually maintain relatively high water content
- Used on higher value agricultural/horticultural crops and in landscapes and nurseries

Advantages

- High application efficiency
- High yield/quality
- Decreased energy requirements
- Reduced salinity hazard
- Adaptable for chemigation
- Reduced weed growth and disease problems
- Can be highly automated

Disadvantages

- High initial cost
- Maintenance requirements (emitter clogging, etc.)
- Restricted plant root development
- Salt accumulation near plants (along the edges of the wetted zone)

Filtration System

Good filtration is essential for proper irrigation system operation and long-term performance. Filters are a must, commonly used to remove sand, silt, precipitated minerals and organic matter for preventing clogging of emitters. Water quality and emmitter specifications will determine the filtration type, level and quantity. Most drip irrigation systems require from 120 to 200 mesh filtration. Irrigation filters will neither remove salt, dissolved solids or other toxic elements, nor will they adjust the water pH. Even if potable water is used, a basic screen filter is required to remove sand and other suspended particles. For good filtration, filters must be cleaned or back flushed, when they become dirty. A two stage filter (primary and secondary) unit may be required for the removal of all impurities.

Primary filters

Hydro cyclone filter

The salient features of this filter are simple and it is an efficient one for filtering or separating sand and other heavier particles like stones from irrigation water. The funnel-shaped body is connected to a connecting tank at the bottom, which collects the separated sand from the water. It is normally connected to the suction line of the pump to reduce wear and tear.

Sand/gravel filter

Sand/gravel filter is mainly used for primary filtration of water drawn from open wells. Water enters through the inlet at top and moves through the media to the bottom. Filtration takes place during this process. Sand media filters provide filtration to 200 mesh, which is necessary to clean surface water and water from open canals for drip irrigation. These water sources pick up a lot of fine grit and organic material, which must be removed before water passes through the drip emitters. Sand media filters are designed to be cleaned through a "back flush" mechanism. The drop in pressure due to the accumulation of the particles is detected with the pressure gauges fitted at the inlet and outlet of the filter. It is then back flushed with water through the sand to remove clay, silt and organic particles by operating appropriate valves . Sand used for filters should be between 16 and 20 mm size to prevent excess back flushing. To assure enough clean water for back flushing, several smaller send media filters are more appropriate than a single large sand media filter.

Secondary filter

Disc filter

This filter is used to remove very fine particles that are present in water. The disc filter comprises of a number of plastic discs, which have very fine radial grooves. These discs are then pressed and held against each other specially designed clamping system. When water flows through discs, the finest particles are filtered.

Screen filter

Screen filter normally consists of stainless steel screen of 200 mesh size (0.075 mm), which is enclosed in a mild steel body. Filtration is achieved by the movement of water through the screen. In all drip irrigation systems, screen filter is needed and will be the last one in the sequence of filters.

Fertigation unit

Fertigation is a method of fertiliser application, where in, fertIliser is combined with irrigation water in the drip irrigation system. Recommended water-soluble fertilisers are mixed with water and passed into the system. In this system, the fertiliser solution is distributed uniformly through irrigation water. The availability of nutrients under this is very high, and hence, the efficiency is more. Liquid fertilisers as well as water-soluble fertilisers are used in drip fertigation, and fertiliser use efficiency can be increased upto 80 to 90%. This unit deliveries nutrients to the plants with water and also applies system maintenance chemicals, such as acid, chlorine or other emmitter cleaners. There are different types of fertigation units available. They include Pump, Tank and Venturi units.

Venturi

This is a very simple and the low-cost devices. A partial vacuum is created in the system, which allows suction of fertilisers into the irrigation system through Venturi action. Vacuum is created by diverting a proportion of water flow from the mainline and passing it through the constriction (Venturi), which increases the velocity of flow, thus creating a drop in pressure. When pressure drops, the fertiliser solution is sucked into the Venturi through a suction pipe from the tank, from where, it enters into the irrigation stream . Even though Simple, Venturi may not mix the fertilisers uniformly with the water, which may result in uneven fertilisers distribution in the field. The suction rate of Venturi is 30 to 120 litres per hour. It is easy to handle and affordable even by small farmers. Venturi is more suitable for small landholdings.

Fertiliser tank

A tank containing solution is connected to the irrigation pipe at the supply point. In this system, part of the irrigation water is diverted from the mainline to flow through a tank, containing the fertiliser in a fluid or granular form. A slight of reduction in pressure is created between the offtake and return pipes of the tank by means of a pressure reducing valve, which causes water from the mainline to flow through the tank, resulting in dilution and flow of the diluted fertiliser into the irrigation stream. With this system, the concentration of fertiliser entering the irrigation water changes continuously with the time, starting from a high to a low concentration. Hence, there will not be uniform fertiliser distribution. Fertiliser tanks are available with 90, 120 and 160 litres capacity.

Fertiliser pump

In this system, the fertiliser solution is held in the non-pressurised tank and it can be injected into the irrigation water at any desired ratio. Therefore, fertiliser availability to each plant is maintained properly. These are piston or diaphragm pumps, which are driven by the water pressure of the irrigation system such that injection rate is proportional to the flow of water in the system. A high degree of control over the fertiliser injection rate is possible and no serious head losses are incurred. But the capital cost is high. Another advantage is that if the flow of water stops, fertiliser injection also automatically stops. This is a perfect equipment for accurate fertigation. Suction rates of pumps vary from 40 to 160 litres per hour.

References

1. Michael A.M., 2012. Irrigation: Theory and Practice, Vikas Publishing House, New Delhi
2. Suresh, R., 2010. Principles of Micro-Irrigation Engineering, Standard Publishers, Distributors, New Delhi.
3. Michael A.M. And Ojha T.P., 2014. Principles of Agricultural Engineering Vol-II 5th Edition. Jain Brothers Publication, New Delhi.
4. Michael A.M., Khepar S.D. and S.K. Sondhi, 2008. Water Well and Pumps, 2nd Edition, Tata Mc-Graw Hill.
5. Sanjay Kumar, Vishal Kumar and Ram Kumar Sahu, 2016. Fundamentals of Agricultural Engineering, Kalyani Publishers, New Delhi.

❏❏❏

Chapter 22

New Generation Insecticides

Dr. Ambily Paul (Asst. Professor) and **Ms. Anju Padmanabhan** (Research Scholar)

Department of Agrl. Entomology Kerala, Agricultural University, College of Agriculture Vellayani, Thiruvananthapuram-695 522

Synthetic pesticides are one of the major agro inputs that are widely used in most sectors of agricultural production to prevent or reduce losses by pests. Food and Agricultural Organization (FAO) defines pesticide as any substance or mixture of substances intended for preventing, destroying or controlling any pest including vectors of human or animal diseases, unwanted species of plants or animals causing harm during or otherwise interfering with the production, processing, storage, transport or marketing of food, agricultural commodities, wood and wood products or animal feed stuffs, which may be administered to animals for the control of insects, arachnids or other pests in or on their bodies. Excessive and indiscriminate use of insecticides has led to problems *viz.*, insecticide resistance, pest resurgence, secondary pest outbreak, accumulation of harmful residues, occupational hazards, environmental pollution, etc.

There is a rising public concern about the potential adverse effects of chemical pesticides on human health, environment and bio-diversity. Awareness and anxiety about risks that residues of pesticides may pose on human health is challenging, especially in the production of vegetables. In 2011, Govt. of Kerala has banned 16 pesticides; belong to red and yellow category in the light of endosulfan issue. The sudden withdrawal of pesticides badly affected farmers, especially when their crops were severely infested with pests. Considering the urgency of the requirement of alternate pesticides, which are safe to the environment and mammals, Kerala Agricultural University came forward with new generation insecticides, which are more selective than conventional insecticides through the project *viz.*, "Development of technologies including alternatives for banned pesticides for the management of pests and diseases of major crops in Kerala funded by RKVY. The chief motto for these developments is to give protection to the crops along with safety to the natural enemies of different pests as a whole safety to environment. Mode of action of new molecules with new target site is different from that of conventional ones. Around 50 new generation molecules are registered in India over the last 5 years belonging to 10 novel classes of insecticides. Merits of new generation insecticides over conventional ones are:

- Narrow spectrum of activity
- Highly selective
- Very low dose (5 to 100 g ai/ha)
- Safe to natural enemies
- Safe to mammals
- New site of action
- No bio-accumulation
- Several of the compounds are effective to both, mites and insects

Important groups of new generation insecticides

1. Neo nicotinoids

Closely related to nicotine in mode of action and in structure, which are developed in 1980s and they replaced organophosphates and carbamates. First commercialised neonicotinoid compound is imidacloprid, which is applied as foliar treatment, seed dresser and as soil treatment and possess contact and systemic activity. Neo nicotinoids are effective against insects having piercing and sucking type of mouth parts. As per IRAC mode of action classification (2018), neonicotinoids comes under the group, nicotinic acetylcholine receptor (nAChR) competitive modulators. They mimic the action of Acetyl choline and interact with Ach binding site of the nicotinic acetyl choline receptor. This interaction causes excitation and eventually paralysis leading to death.

Classification of neo nicotinoids

1. Chloronicotinyls e.g. - Imidacloprid, Acetamiprid
2. Thianicotinyls e.g. - Thiamethoxam, Clothianidin
3. Furanicotinyl e.g. - Dinotefuran

All neonotinoids, especially imidacloprid and thiamethoxam are toxic to honey bees. Several reports are available regarding the development of insecticide resistance in insects against imidacloprid.

2. Nereistoxin Analogues

It is an analogue or propesticide of the natural product *nereistoxin,* which is a potent insect neurotoxin isolated from the marine annelid, *Lumbriconereis heteropoda.* These are nicotinic acetyl choline receptor channel blockers.

Synthetic analogues of nereistoxin are bensultap, cartap hydrochloride, thiocyclam, thiosultap-sodium. Cartap hydrochloride is the most popular one and effective against coleopteran and lepidoptera and act as contact and stomach poison.

Nereistoxin is different from neo nicotinoids in the sense that it is blocking cholinergic transmission resulting in the paralysis and death of insects without convulsive symptoms, while neo nicotinoids mimic the action of acetyl choline.

3. Spinosyns

The active ingredient of this group was obtained from a naturally occurring soil bacterium (actinomycetes), *Saccharopolyspora spinosa* . Spinosad contains a mixture of two spinosoids, spinosyn A, the major component, and spinosyn D,the minor component.

They are called Nicotinic acetylcholine receptor allosteric modulators. The spinosyns and spinosad have a novel mode of action, primarily targetting binding sites on nicotinic acetylcholine receptors of the insect nervous system that is distinct from those at which other insecticides have their activity. Spinosad binding leads to disruption of acetylcholine neurotransmission. It has high efficacy, a broad insect pest spectrum, low mammalian toxicity, and a good environmental profile.

4. Avermectins

Derivatives of macrocyclic lactones and these are naturally occurring compounds isolated from soil actinomycetes *Streptomyces avermitilis.* Natural and synthetic derivatives of avermectin includes abamectin, ivermectin and emamectin benzoate.

Abamectin (Mixture of 80% avermectin B1a + 20% avermectin B1b). Popular and widely used molecule under the group is emamectin benzoate.

They act as glutamate gated chloride channel (GluCl) allosteric modulators. The avermectins block the transmission of electrical activity in invertebrate nerve and muscle cells mostly by enhancing the effects of glutamate at the invertebrate-specific glutamate-gated chloride channel, with minor effects on gamma-aminobutyric acid receptor. This causes influx of chloride ion into cells, leading to hyperpolarisation.

5. Phenyl pyrazoles

Broad spectrum activity and act as Gamma Amino Butric Acid (GABA) gated chloride channel blocker. Important compounds under this group are ethiprole, fipronil.

Fipronil has a broad spectrum of activity against insects. It is effective at low field application rates against insects that are resistant to pyrethroids, organophosphates and carbamates. Highly toxic to many aquatic species. Fipronil disrupts the insect central nervous system by blocking GABA.

6. Diamides

Synthetic derivatives of the alkaloid Ryanodine obtained from *Ryania speciosa, R. tomentosa* and *R. sagotina.* The water soluble plant extracts of *Ryania* has been used as an insecticide for about 50 years. The extracts contains several structurally related ryanodine *viz.,* ryanodine, 10-(0-methyl)-ryanodine, 9, 21-dehydro ryanodine and ryanodol. Diamide has broad spectrum activity against lepidopterans.

The primary site of action of diamide is ryanodine receptor modulators. Diamides activate ryanodine sensitive intra-cellular calcium release channels (RyRs) in insects. They bind in the pore region of the receptor and lock the channel in the open state and release the Ca^{2+} ions from internal stores into the cytoplasm. Depletion of Ca^{2+} stores results in paralysis and death of an insect. Excellent selectivity favours beneficial insects and selectivity to mammals.

Important compounds under this group are: chlorantraniliprole, flubendiamide, cyantraniliprole and tetraniliprole. All these compounds come under the toxicity class "green".

7. Tetronic and tetramic acid derivatives

These compounds are launched in 2002-04 in the world market. It has contact with slight translaminar activity, acaricidal and insecticidal activity. Mode of action is inhibitors of acetyl CoA carboxylase. Block the fat synthesis, which ultimately causes the target pests to dry out and die because it losses water balance due to the inhibition of lipid bio-synthesis. Important compounds come under this group are spirodiclofen, spiromesifen, spirotetramat.

8. Insecticide mixtures

Insecticide mixtures are the combinations of two or more pesticides having different mode of action into a single spray solution, which entails exposing individual in an arthropod pest population to each pesticide simultaneously. The primary benefits of insecticide mixtures are delay in development of insecticide resistance, less number of applications, less labour cost, chemical cost, less dosage when compared to single insecticides and control of pests in a broad range. Total of 31 products registered for agricultural use as per CIBRC website (www.cibrc.nic.in). Among this, around 15 numbers are the combination of new generation insecticides.

e.g.:

1. Chlorantraniliprole 8.8% w/w + Thiamethoxam 17.5% w/w SC
2. Flubendiamide 4% + Buprofezin 20% w/w SC
3. Fipronil 40% + Imidacloprid 40% WG.

List of new generation insecticides recommended against pests of crops in Kerala						
Sl. No.	**Insecticide recommended**	**Crop/ pest**	**Dosage (g ai / ha)**	**Field Dose (mg or ml L^{-1}**	**Colour code (Toxicity class) of chemical**	**Trade name**
1.	Chlorantraniliprole 0.4 % G	Rice stem borer, gall midge and leaf folder	40	---	Green	Ferterra
2.	Chlorantraniliprole 18.5 % SC	Rice stem borer and leaf folder	30	0.3 ml/L	Green	Coragen
3.	Fipronil 0.3% G	Rice gall midge	50	----	Yellow	Sonic, Regent, Biltop, Sarget
4.	Carbosulfan 6 %G	Rice stem borer, gall midge and leaf folder	1000	----	Yellow	Sherief, Marshal
5.	Buprofezin 25 % SC	Brown plant hopper	200	1.5 ml/L	Blue	Applaud
6.	Spiromesifen 22.9 % SC	Chilli mite & thrips, Brinjal-white fly, Jassids and Mites	96	0.8 ml/L	Blue	Oberon,
7.	Fenpyroximate 5% EC	Chilli mite and thrips	15	0.6 ml/L	Yellow	Mitigate, Sedna
8.	Thiamethoxam 25 % WG	Brinjal- white fly, jassids and mite Bhindi - jassids and whiteflies Chilli - mite & thrips	50	0.2 g/ L	Blue	Actara, Suckgan,
9.	Diafenthiuron 50 % WP	Brinjal -white fly, jassids and mite	300	1.2 g/ L	Blue	Pegasus, Polo,
10.	Spiromesifen 22.9 % SC	Brinjal -white fly, jassids and mite	96	0.8 ml/L	Blue	Oberon
11.	Chlorantraniliprole 18.5 % SC	Cowpea-pod borers, leaf eating caterpillars Snakegourd-pumpkin caterpillar Brinjal, Bhindi-Shoot & fruit borer Cowpea- Epilachna beetle	30	0.3 ml/ L	Green	Coragen
12.	Emamectin benzoate 5 % SG	Brinjal, Bhindi - Shoot and fruit borer	500	0.3 g/ l	Yellow	Proclaim
13.	Imidacloprid 17.8 % SL	Bhindi - jassids and whiteflies	30	0.3 ml/ L	Yellow	Tatamida, confidor

		Cowpea- aphid and pod bugs				
14.	Acetamiprid 20 % SP	Brinjal- - white fly, jassids and mite Cowpea- pod bugs	15	0.2 g/ L	Yellow	Tatamanic , pride
15.	Acephate 75 % SP	Brinjal - white fly, jassids and mite	500	1 g/L	Blue	Asataf, Starthene, Lancer
16.	Thiacloprid 21.7 % SC	Cowpea- aphid and pod bugs	30	0.3 ml/ L	Yellow	Calypso, Alanto
17.	Thiodiarb 75 % WP	Bittergourd- Epilachna beetle	470	1.25 g/L	Yellow	Larvin, spiro,
18.	Indoxacarb 4.5 % SC	Bittergourd- Epilachna beetle Brinjal, Bhindi - Shoot and fruit borer	50	0.8 ml/ L	Yellow	Avaunt, hindoxa
19.	Spinosad 45 % SC	Brinjal, Bhindi- Shoot and fruit borer	75	0.3 ml/ L	Blue	Tracer

References

GOK [Government of Kerala], 2018. *GOK Plan project,* Research on new molecules of plant protection chemicals. Government of Kerala, Department of Entomology, College of Agriculture, Vellayani, 22p.

Gour, T. B. and Sridevi, D. 2017. Chemistry, Toxicity and mode of action of Insecticides, Kalyani Publishers, India. 316p.

KAU [Kerala Agricultural University], 2016. *Package of Practices Recommendation: Crops* (15th Ed.). Kerala Agricultural University, Thrissur, 392p.

□□□

Chapter 23

Microbial Inoculants for Crop Disease Management and Nutrition

Dr. K.S. Meenakumari (Professor & Head), **P. Subha, Bindhu R. Raj** and **S. Vigi** (Research Scholars)

Kerala Agricultural University, College of Agriculture Vellayani, Thiruvananthapuram-695 522

The increasing reports on the development of resistance among pathogens against pesticides, pollution of water and foodstuff and related environmental problems have led to an upsurge in the interest in biological control of plant diseases. Decades of laboratory and field experiments have resulted in the successful launching of more than 50 commercial formulations comprising of fungi and bacteria in the world market for the biological control of plant pathogens. Biological control of plant pathogens can be broadly defined as the reduction of inoculum density or disease producing activities of a pathogen or parasite in its active or dormant state by one or more organisms, accomplished naturally or through manipulation of environment, host or antagonist or by mass introduction of one or more antagonists.

Fungal antagonists, such as *T. viride, T. harzianum, T. virens, T.longibrachyatum* and bacterial antagonist Fluorescent pesudo-monads, *Bacillus subtilis* have been widely exploited and are available in the international market.Fluorescent pseudomonads and species of *Bacillus* are effective in the control of most of the phytopathogenic bacteria and fungi. This includes species of *Xanthomonas, Ralstonia* and fungal pathogens, *Pythium, Phytophthora, Rhizoctonia, Alternaria, Colletotrichum, Fusarium*, Sclerotium, etc. *Trichoderma* spp. are among the most promising biocontrol agents found highly successful for biocontrol of wide range of plant pathogens including *Armillaria mellea, Pythium* spp., *Phytophthora* spp., *R. solani, Sclerotium rolfsii, Sclerotinia* spp., *Fusarium* spp., *Heterobasidium* spp., *Aphanomyces* spp., etc.

AMF as a bio-control agent

Arbuscular mycorrhizae are now well-recognized as a bio-fertilizer-cum-deterrent to root pathogens and widely used for biological control of root diseases of crop plants. A wide variety of reports in recent years have indicated that host plants previously inoculated with a mycorrhizal fungal symbiont exhibits increased tolerance to a range of root diseases. This includes a spectrum of diseases caused by species of *Fusarium, Phytophthora, Pythium, Rhizoctonia, Thielaviopsis, Macrophomina, Verticillium, Sclerotium, Cylindrocladium* and *Olpidium.* Successful suppression of root-knot nematode and burrowing nematode infestation in a variety of crop plants was obtained consequent to pre-inoculation with appropriate mycorrhizal fungus.

Mode of action of bio-control agents

There are number of ways by which antagonistic organisms suppress growth of pathogens and control disease. Their action could be through: (1) competition for nutrients, (2) parasitism, (3) production of anti-biotics and volatile compounds (4) lytic enzymes, (5) antagonistic proteins and (6) induction of host resistance.

Fluorescent pesudomonads compete with pathogens for iron. They produce iron-sequestering siderophores, which decrease iron and inhibit the growth of pathogens. Siderophore like pyochelin, pyoverdine, pseudobactin, ferribactin, ferrichrome and ferroxamine are secreted by psuedomonads. Iron starvation prevents the germination of fungal spores. Some organisms like *Trichoderma* spp. directly parasitize on many soil borne fungal pathogens (*Pythium, Phytophthora, Rhizoctonia, Fusarium, Sclerotium*). The hyphae of parasitic type *Trichoderma* grow towards and coil around the pathogen hyphae and partial degradation of the cell wall takes place . However, the microbial action against fungal and bacterial pathogens is usually through the production of toxic substances. *T.viride* produces antibiotics like trichodermin, dermadin, and trichoviridin. *T. longibrachiatum* produces trichodermin and trichobrachin. *B. subtilis* produces antibiotics iturin, bacilycia and fengacin, which inhibit *Rhizoctonia, Botrytis* and *Xanthomonas. Pseudomonas fluorescens* produces antibiotics pyoluteorin, phenazines, oomycin, tropolone, pycocyanin and 2, 4-diacetyl 1-phloroglucinol having broad-spectrum activity against pathogenic fungi and bacteria . Strains of *T. harzianum* were found to produce two volatile metabolites 6-n-pentenyl-2H-pyran-2-one and 6n-pentyl-2H-pyran-2-one. *T. viride* produced lactones, alcohols and terpene derivatives of pyran. These volatile substances inhibited the mycelia of a spectrum of pathogens. The extra cellular enzymes like chitinase and glucanase are produced by *Trichoderma* spp. These enzymes degrade the cell wall of the pathogen. *B.subtilis* produced antagonistic proteins strongly inhibitory to seed borne fungi and bacteria.

Some bio-control agents induces resistance in host plant. An example is the use of arbuscular mycorrhizal fungi (AMF). AMF inoculation induces resistance against foot root of black pepper, rhizome rot of ginger, Azhukal disease of cardamom and root-knot nematode infestation in most of the crop plants. Various mechanisms have been proposed for AMF mediated resistance to root pathogens. Improved nutrition of mycorrhizal plants, particularly the P uptake, enables to resist the root diseases. The competition for host photosynthate and infection site in the root offered by mycorrhiza affects the successful infection and establishment of the pathogen in the root. Anatomical and morphological changes in the root system, such as increased lignification, root branching, meristematic and nuclear activities of root cells and the microbial changes in the mycorrhizosphere, particularly the increased activity of antagonistic organisms, due to AMF colonization also affect the infection and development of pathogens. Stimulation of plant defence mechanisms has also been noticed in mycorrhizal symbiosis. Accumulation of amino acids arginine and citrulline, higher activity of chitinase, glucanase, peroxidase, elicit production of PR proteins and higher levels of proline rich glycoproteins, phytoalexins and phenolics are some factors involved in plant defense in relation to AM formation.

Bio-fertilizers for crop nutrition

This microbial diversity could be exploited for augmenting the supply of major nutrients, such as N and P in crop production. The critical need is to enhance the bio-availability of nutrients for crop production. Efficient micro-organisms are available in the soil which facilitate plant growth through nitrogen fixation, solubilization of fixed forms of phosphate, production of phytohormones, etc. that can modulate plant growth and development. The role of bio-fertilizers in sustainable agriculture assumes special significance and it has been clearly demonstrated that this technology is a powerful tool for enriching soil fertility that improves crop yield.

Of the different bio-fertilizer organisms, diazotrophs like *Rhizobium*, *Azospirillum* and *Azotobacter* supply the nutrient nitrogen.

Rhizobium is an important symbiotic nitrogen fixing system which accounts for 40 per cent of the global nitrogen fixation. *Azotobacter* and *Azospirillum* are free living and associative nitrogen fixing bacteria, respectively used for different crops. Potential nitrogen fixers in the wetland rice include cyanobacteria and Azolla. Apart from nitrogen fixers, some genera of bacteria and fungi are capable of solubilizing insoluble forms of phosphorus in the soil and play significant role in P nutrition of crop plants.

Bio-fertilizers for Nitrogen Nutrition

Nitrogen is an essential plant nutrient, which is the most commonly deficient element responsible for reduced agricultural yield throughout the world. Even though atmospheric air contains 78 per cent nitrogen, none of the living organisms other than a group of microbes can directly absorb the elemental nitrogen. Bacteria like *Rhizobium, Azotobacter, Azospirillum,* blue green algae and azolla are capable of absorbing the elemental nitrogen and fixing it through biological nitrogen fixation.

Legume – *Rhizobium* Symbiosis

The global nitrogen fixation is estimated to be 175 million tonnes per year (Burns and Hardy, 1975). Legume - *Rhizobium* symbiosis alone contributes 40 per cent, which accounts for atleast 70 million tonnes per year. This symbiosis is infact a gift of nature to mankind, which meets nearly 60-80 per cent of the nitrogen requirement of the host legume. Besides, the cultivation of legumes will also result in enrichment of soil with fixed nitrogen. The success of legume- *Rhizobium* symbiosis, however, is dependent on several biotic as well as abiotic factors. Ecological factors like soil acidity, alkalinity, application of agricultural chemicals and presence of ineffective strains, antagonistic and effective strains influence establishment of introduced strains. Legume plants having effective root nodules can meet not only its own nitrogen requirement but also enrich the soil nitrogen content and thereby improve soil fertility. The practice of seed inoculation with *Rhizobium* is popular among farmers which increases both, N_2 fixation as well as yield of inoculated plants. The N_2 fixation by different rhizobia vary widely from 30-300 kg $ha^{-1.}$ The nitrogen fixation by different legumes were estimated to be 100-300 in alfalfa, 102-150 in clover, 37-196 in clusterbean, 112-152 in peanut, 68-200 in pigeon pea, 49-130 in soyabean and 36-63 kg ha^{-1} in chick pea.

Green manures could be fitted into many cropping systems to derive benefit of legume - *Rhizobium* symbiosis for N nutrition. Among the green manures, *Indigofera tinctoria, Calapogonium* spp., *Crotalaria, Clitoria, Desmanthus* and *Macroptilium* have substantial ability to fix atmospheric N_2.

Azorhizobium caulinodans, which induces root and stem nodules in tropical legume, *Sesbania rostrata* is especially interesting, since in addition to forming nodules on emerging lateral roots, it is able to fix nitrogen in the free-living state.

Bio-fertilizers for phosphorus nutrition

Phosphorus is the "Master Key" element not only in plant metabolism but also in soil micro-biological processes. It is second only to nitrogen as an essential nutrient for plant growth and development. Various metabolic activities occurring in the plant are also related to P supply. P is added to soil in the form of phosphatic fertilizers, part of which is utilized by the plant and the remaining part is fixed as insoluble forms.

Phosphorus Solublizing Micro-organisms (PSM)

Any technology that aids in the rapid availability of P from fixed or insoluble forms in the soil, will be of much field application. There has been long standing interest to manipulate soil micro-organisms to improve P nutrition of plants. The concept of using soil micro-organisms to improve mobilization of poorly available and fixed forms of soil P is not new. A wide spectrum of soil micro-organisms are capable of solubilizing precipitated forms of P or mineralizes organic P. Species of *Bacillus, Pseudomonas, Micrococcus, Flavobacterium, Aspergillus, Penicillium, Mucor, Fusarium, Candida,* etc. are reported to be pre-demoninant. These organisms are commonly associated with the rhizosphere and when inoculated into plants result in improved growth and P nutrition. They have the capacity to solubilize chemically fixed soil P, rock phosphate and even organic P due to acid production and enzymatic activity.

Significant increase in yield of rice, wheat and maize has been obtained, reviewed the status of microbial inoculant technology and emphasized the need for evolving efficient strains adapted to the local conditions to exploit the full potential of the technology for crop production. PSM alongwith acid production produces the phosphatic enzymes, which cause solubilization of P in aquatic environment. The production of chelating substances H_2S, CO_2, mineral acids, siderophores are also involved in P solubilization. PSM not only releases plant available P but also produces biologically active substances like IAA, Gibberellins, Cytokinins, etc., which are known for their growth promotion activity in plants. Beneficial effect of PSM in increasing crop yield as well as in reducing the requirement of phosphatic fertilizers has already been established. A saving of 50 per cent of recommended P_2O_5 is possible with the inoculation of phospho-bacterium.

Arbuscular mycorrhizal fungi

Mycorrhizal association is the mutualistic symbiosis occurring between certain soil fungi and roots of higher plants, which is now well-recognized as a bio-fertilizer. The beneficial effects of AMF associations in the host nutrition, especially in the uptake of 'P' is well documented. Mycorrhizal colonization enhances plant growth by increasing nutrient uptake either by increasing the absorbing surface and mobilizing sparingly available nutrient sources or by secretion of ectoenzymes. AM plants absorb and accumulate more 'P' than non-AM plants either due to more efficient translocation coupled with better exploration of soluble 'P' by the fungal hyphae or due to solublization of insoluble 'P' by the AMF. The role of AMF in the uptake of other nutrients viz., Cu, Zn, Ca, K, Fe, Mn is also well documented. Further, AM association increases the uptake of water , reduce stress responses of plants to toxicity and drought, improve soil texture and reduce soil erosion and enhances revegetation of degraded soils like mine soils.Increased yield through enhanced uptake of nutrients, especially P has been obtained in several crops like cowpea, cassava, ginger, cardamom, cashew, etc.

Bio-fertilizers for potassium nutrition

Potassium is one of the essential macro-nutrient and the most abundantly absorbed cation and the element plays important role in growth and development of plants. Micro-organisms play key role in potassium availability. *Bacillus mucilaginous* and *Frateuria aurantia* are important potassium solublizers. Most of the potassium in soils exists in the form of silicate minerals. Potassium solublizing bacteria are capable of solubilizing rock potassium, mineral powder, such as mica, illite and orthoclases through production and excretion of organic acids.

□□□

Chapter 24

Rice - Lifeline of Kerala

Dr. Beena R. (Asst. Professor), and
Chennamsetti Lakshmi Naga Manikanta (Research Scholar)

Department of Plant Physiology, Kerala Agricultural University, College of Agriculture Vellayani, Thiruvananthapuram-695 522

Cultivation of rice has been practiced in Kerala for several centuries. Apart from being as a prime food source for 90% population, it has its noteworthy mark in the culture of Kerala. Rice cultivation and varieties used in Kerala assume greater importance due to the diversity in rice growing situations and availability of naturally occurring donors to diverse stress situations. This vast diversity of land races and the occurrence and characteristics of wild races lead to the theory that Kerala may be probable centers of origin of *Oryza sativa* (L.).

Climate and Season

State is blessed with typical tropical-monsoon with seasonally excessive rainfall and hot summer, this follows an exception i.e., southern tip of state Thiruvananthapuram, is of tropical - savannah with seasonally dry and hot weather. Rice growing seasons in Kerala were broadly classified into three:

1. Virippu (first crop/autumn); April-May to Sept-Oct.
2. Mundakan (second crop/winter); Sept-Oct to Dec-Jan
3. Puncha (third crop/summer); Dec-Jan to Mar-Apr.

Rainfall and water source

Kerala has an ample source of water, it has 44 rivers, 27 backwaters (mostly in the form of lakes and ocean inlets), 7 Lagoons and 18681 ponds and over 30 lakh wells. To keep these sources alive,

Kerala state is having quite high average level of annual rainfall, being in the neighbourhood of about 3600 mm. Kerala gets rainfall from both the monsoon. Palakkad receives minimum rainfall; whereas Idukki district has the highest rainfall of 3000 mm.

Rice Ecosystems of Kerala

The state of Kerala, a narrow strip of land approximately 600 km long varying in breadth from 30 to 120 km; lying largely on the western aspects of Western Ghats is conventionally distinguished into three climatically separate regions, namely the eastern highlands; the central mid lands and the Western lowlands. The state's landmass is distinguished into 13 agro climatic zones based on altitude, rainfall pattern, soil type and topography . Rice is a very versatile crop allowing it to cultivate in distinct agro ecosystem across the state, starting with lands below sea level as in the case of Pokkali fields to rice paddies in the high ranges of 800-1500 m MSL. Out of these, important rice growing tracts are as follows:

Kole Paddy Fields

Trissur and Malappuram district of Kerala.The land is linked with sea through drainage channels & backwaters. The total extent under paddy is 18602 Ha, which is 2.35% of the gross rice cultivation area in the state. Kole wetlands are present above MSL; so that intrusion of seawater is checked enabling the lands to cultivate Rice.

Paddy is cultivated only only during months of December to April in karuvannur and kecheri basins. At present, only one crop is raised & the rest of year, the tracts gets flooded. Kol wetlands were under rice cultivation for the past 200 years. Large quantities of nutrients rich alluvial soils get deposited during the process of inundation making the wetland highly fertile for paddy cultivation. Human interference into this wetland dates back to the early 18th century.

Important varieties suitable to this tract are:

SEASON : MUNDAKAN

Extra short duration: Harswa.

Short duration : Karthika, Makom, Aruna, Jyothi, Onam, Bhagya, MattaTriveni, Revathi, Annapoorna, Remanika.

Medium duration : Pavizham, Remya, Kanakom, Jaya, Rajini, Pavithra, Uma, Aswathy, Sabari.

Pokkali tract

Pokkali has been cultivated in the coastal fields of Kerala, bordering the Arabian sea for the last 900 years and is one of the traditional variety of paddy in the south Asian continent. Pokkali which alternates between paddy and prawns is practiced in the wetlands of Alappuzha, Thrissur and Ernakulum districts. Marshy tracts lie near the mouth of the rivers and are close to the sea. Therefore they are prone to flooding and salinity. The nature of the soil here is also highly acidic & has high element toxic content, making the areas uncongenial for paddy cultivation, but the traditional method has been proved by the scientists as the best method to prepare the fields for Pokkali because of the larger surface area and it also enhances the integration of toxic contents of the soil with rainwater. Pokkali paddy, which to an extent is saline, flood resistant and acid resistant, is the wonder crop cultivated in these regions.

Pokkali is the oldest variety of rice in Kerala, which has a tradition of at least 3000 years in rice cultivation. It is also among the oldest known crops cultivated by organic farming methodologies in the world. In rice cultivation, the Pokkali variant has remained a favourite among many farmers and now attained global recognition as well as exclusive cultivation and distribution rights to its Kerala growers.

Paddy cultivation is practiced in Pokkali tract during the low saline phase and prawn farming is practiced during the high saline phase . Starts cultivating of soil from auspicious Malayalam New Year day, every year.

Varieties suited: First crop: Vyttila 1, Vyttila 2, Vyttila 3, Vyttila 4, Vyttila 5, Vyttila 6, Vyttila 7, Vyttila 8, Vyttila 9.

Kuttanad Tract

Low lying areas in central Kerala, South of Kochi spread out in the districts of Alappuzha, Kottayam, Pathanamthitta in the taluks of Chertallai, Ampalappuzha, Kuttanad, Karthikapally, Thiruvalla, Chengannur and Mavelikkara. Total area under Kuttanad is 1100 Sq. Km, out of which 304 Sq. Km is 1 meter above sea and 500 Sq. Km is 0.6 to 2.2 m below sea level, 40% of Kuttanad is engaged in paddy cultivation. Kuttanad is divided into Karapadam (83,000 Ha.), Kayal lands (13,000 Ha.) Kari soils (9,000 Ha.). Karapadam soils are river borne alluvial soils which are silty clay, moderately acidic high salt content, and a fair amount of decomposition organic matter. Kayal lands are reclaimed beds of Vembanad. There are 1167, 'Padasekharams' with an area of 2 to 1000 Ha. In Kuttanad, the bunds around the Padasekharams protect the field by preventing the water from outside entering the fields. Cultivation happens during North-East monsoon. Two crops, namely : Virippu-middle of May to end of June and Puncha-end of November to end of March.

Kaippad Rice Paddies

Kaippad rice paddies has been cultivated in Ezhome, Narikkode, Kottila, Pallikara, Vettukaadu, Muttil, Muttam, Chelikkalkaduvu, Perumba and Kunjimangalam areas of Kannur districts between Pazhayangadi and Talimparamba areas on the banks of Pazhayangadi river.

Capture of fishes will be practiced during Oct to April. From April onwards, Kaippadu areas will be kept for drying. After first rains, germinated paddy seeds are broadcasted in the field.

Potta cultivation: Soil mounds are made using heaps of soil in the field called "Potta", which will end by the end of April.

Kaippad varieties : First Crop : Ezhome 1, Ezhome 2, Ezhome 3, Ezhome 4.

Research Wings of Rice in Kerala

1. Regional Agricultural Research Station, Pattambi

Established as paddy breeding station by British in 1927, made tremendous contribution to betterment of rice production. In 1930, named as Agricultural Research Station and in 1956, became a part of Kerala state.

First high yielding hybrid derivative variety of rice of South-East Asia. Annapoorna (Ptb-35) in 1966.

Research contribution are continued with useful outfits for the crop production sector in general & rice production in particular. So far release, 62 rice varieties (Ptb 1-Ptb 62) was released.

2. Rice Research Station, Moncompu

Centre is exclusively to deal with research on rice in Kuttanad ecosystem. A unit of All India Co-ordinated Rice Improvement Programme is functioning here. Centre has so far released 12 rice cultivars. These are Mo1, Mo2, Mo3, Bhadra, Asha, Pavizham, Karthika, Aruna, Makom, Remya, Kanakam and Ranjini and are well accepted by the farmers. Another seven pre-release cultures were proposed for release. They are: Pavithra, Panchami, Remanika, Uma, Revathy, Karishma and Krishnanjana.

3. Rice Research Station, Vytilla

Situated in a representative site in the centre of the Pokkali tract. This station started functioning during the year 1958. Taking into account, the importance of fish- prawn culture during the saline phase, a unit for fisheries research was established during the year 1976.

Rice in *Pokkali* (Saline) areas and rice-fish farming system are the lead functions.

4. Onattukara Regional Agricultural Research Station, Kayamkulam

Established in the year 1937, under the erstwhile University of Travancore. In April, 2000, this station was upgraded to the status of RARS (Onattukara Region).

Lead functions are on rice in *Onattukara,* coconut, annual oilseeds and pulses. Verification functions are on vegetables in homesteads, mushrooms, etc. The station is implementing the Comprehensive Coconut Care Programme.

5. Integrated Farming Systems Research Station, Karamana

The centre, formerly known as the Model Agronomic Research Station, was established in 1955. New schemes and projects were started from 1968 onwards under the All India Co-ordinated Agronomic Research Project (AICARP) of the ICAR. From October, 1983 onwards, the station was upgraded as the HQ of the AICARP in Kerala.

Lead functions are rice-based cropping system and verification functions are agro-techniques (rice), integrated farming systems with rice/fish/poultry/cattle. From 1986, a unit of All India Co-ordinated Agronomic Research Project is functioning here.

Indicators for Rice Cultivation in Kerala

The parameters given below are indicators and actual rates and parameters can vary based on location and method of cultivation.

Land preparation –mainly tractor and power tiller are used along with :

- Tractor + Cultivator
- Tractor + Cage wheel
- Tractor+ Mouldboard plough
- Tractor + Disc plough
- Tractor+ Disc harrow
- Tractor + Puddler

Average Cost

- For wetland ploughing – Rs.900 / hr
- For dryland ploughing- Rs. 750/ hr
- For rotovator- Rs.900/ hr

Lime application

- Lime @ Rs. 6 /kg (subsidized rates)

Nursery Area

- 1/10th of the land area
- Mat nursery should be prepared for transplanter.

Seed rate

- For transplanting 30kg/ acre
- For broadcasting: 40 kg/ acre
- For dibbling: 40 kg/acre
- Seed pellets : 6 kg/acre

Seed Treatment

- *Pseudomonas fluorescence:* 10g / kg of seeds
- Carbendazim: 2g / kg of seeds
- Soak seeds in $CuSo_4$ (0.25 %), ZnSo4 (1%) for 24 hr

Transplanting / Sowing

- 18-25 days old seedlings are used for transplanting
- Manual transplanting - Rs. 4000-4500/acre
- Mechanized transplanting-Rs. 4000/acre (Yanjisakthi)
- Seed drill/Seed drum are also used.

A **rice transplanter** is a specialized transplanter fitted to transplant rice seedlings onto paddy field. Mainly two types of rice transplanter i.e., riding type and walking type. Riding type is power driven and can usually transplant six lines in one pass. On the other hand, walking type is manually driven and can usually transplant four lines in one pass.

A common rice transplanter comprises of:

- A seedling tray like a shed roof on which mat type rice nursery is set;
- A seedling tray shifter that shifts the seedling tray like a carriage of typewriters; and plural pickup forks that pick up a seedling from mat type nursery on the seedling tray and put the seedling into the earth, as if the seedling were taken between human fingers.

Machine transplanting using rice transplanters requires considerably less time and labour than manual transplanting. It increases the approximate area that a person can plant from 700 to 10,000 square metres per day.

Walking type rice transplanter

Riding type transplanter

Spacing

- 10 X15 cm^2, 15 X 15 cm^2

Nutrient Management

- Apply organic manure in the form of farmyard manure or compost or green leaf @ of 5 t/ha and incorporate into the soil while ploughing. The entire quantity of phosphatic fertilizers may be applied alongwith the organic manures.
- Cowpea may be raised as an inter-crop in dry seeded lowland (semi-dry) rice by sowing 12.5 kg seed/ha alongwith rice to serve as a source of green manure. When the rice field gets submerged with the onset of south-west monsoon, cowpea at the age of about six weeks and at active vegetative stage decays and gets self-incorporated in the soil adding substantial quantity of green manure. Such a system of concurrent growing of cowpea also reduces weed pressure in semi-dry rice.
- In general, addition of lime is absolutely necessary, when the pH is lower than 5.5 and it is advisable, when pH varies between 5.5 and 6.5.
- For direct seeded crops during the first season, apply lime @ 600 kg/ha in two split doses, the first dose of 350 kg/ha as basal dressing at the time of first ploughing and the second dose of 250 kg/ha as top dressing about one month after sowing.

- For transplanted crop, apply lime @ 600 kg/ha in two split doses, 350 kg/ha as basal dressing and 250 kg/ha as top dressing about one month after transplanting.
- A time lag of one week should be given between application of lime and fertilizers. For top dressing, lime may be applied one week prior to the application of fertilizers.

Fertilizer recommendation for rice Kg/ha

Kind of land / region	*Variety*	*N*	*P_2O_5*	*K_2O*
Uplands (*modan*)	PTB 28, 29, 30	40	20	30
-do-	High yielding short duration varieties	60	30	30
Wet lands (All regions)	-do-	70	35	35
-do-	High yielding medium duration varieties	90	45	45
-do-	Local varieties	40	20	20
-do-	H4	70	45	45
-do-	Mahsuri	50	25	25
Kole lands	*Short duration varieties	90	35	45
	*Medium duration high yielding varieties	110	45	45
Kattukampal and Ponnani kole lands	*Medium duration high yielding varieties	110	45	55

- For *modan* cultivation (upland crop) and direct seeded crop in wet lands, apply nitrogen in three equal split doses, first as basal dressing, second at tillering stage (three weeks after seeding) and the third at panicle initiation stage (about thirty days before flowering).
- Apply the full dose of P_2O_5 at the time of land preparation as basal dressing. Apply K_2O either in a single dose as basal or in two split doses half as basal and half at the panicle initiation stage.
- The general principle to be followed is that in light soils as well as in soils with high leaching, N may be applied in three or four split doses according to the duration of the variety.

- During the first crop season, when basal application of N is not possible due to incessant rains, basal, basal dose can be shifted to 15 days after transplanting.
- For dry sown Mahsuri, the fertilizer dose of 50:25:25 kg/ha $N:P_2O5:K_2O$ is sufficient. Nitrogen dose may be given in three equal splits at basal and 45 and 85 days after sowing.
- For pre-planting application, apply the fertilizers at the final ploughing. In areas, where availability of water is assured, temporarily draining the field one day prior to application and re-flooding after twelve hours is recommended for top dressing of fertilizer.
- For increasing the efficiency of urea for top dressing, mix urea with six times its weight of slightly moist soil and apply to the field 24-28 hours after mixing. Oilseed cakes, such as *punna* and *neem* cakes can also be mixed with urea (1 part of oil cake + 5 parts of urea by weight) for increasing fertilizer efficiency. This method is particularly useful for basal application of N. Under special conditions of drought and water logging, apply N as foliar spray. Urea may be applied as a low volume spray at 15% concentration using power sprayer or at 5% concentration using a high volume sprayer, the quantity applied in one application being limited to 15 kg/ha.
- Whenever carbofuran application is needed at around 20 days after planting, it would be advantageous to mix 10 kg N in the form of urea. Carbofuran should be used @ 0.75 kg ai/ha.
- When zinc deficiency is noticed, apply zinc sulphate @ 20 kg/ha. Early stages of zinc deficiency are evidenced by interveinal chlorosis, bleaching of midribs and light yellow colouration of the leaf. Older leaves develop brown rusty spots and are extremely brittle. Zinc sulphate and potash should not be applied on the same day.
- Split application of water-soluble phosphatic fertilizers in two equal splits as basal and at maximum tillering stage is effective in giving higher grain and straw yield than the full dose as basal dressing under certain situations.
- When the soil has less than 10 ppm of calcium chloride extractable sulphur or 15 ppm of phosphate extractable sulphur, substitute urea with ammonium phosphate sulphate to correct

sulphur deficiency. For medium duration rice grown in brown hydromorphic soils ammonium phosphate sulphate may be used to supply 25 kg sulphur per hectare to protect the crop from sulphur deficiency (*ad hoc* recommendations).

Weeding

Conoweeder - Single rotor, double rotor (Rs.1200)

Power operated conoweeder.

Spraying

- Knapsack sprayer (manual)
- Power operated back mounted sprayer
- Rocker sprayer (both, manual and power operated)
- Duster
- Tractor mounted sprayer

Harvester

- Combine harvester with 2 m and 5 m cutting width
- Smaller one has low storage capacity
- For belt type: Rs 2200/ hr
- For wheel type : Rs 2500/ hr

Reaper

- KAMCO have 120 cm cutting width (walking behind type)
- Tractor mounted reaper
- Brush cutter

Thresher

- Hold on drum type - small and have 1 m width (for small land holding)
- Feeding type : High efficiency.

Stages of fertilizer application

Kind of land / region	Variety	Stages of application									Remarks
		N			P			K			
		BA **	AT	API	BA	AT	API	BA **	AT	API	** For direct seeded crop, one week after sowing
Uplands	PTB 28, 29, 30	1/3	1/3	1/3	Full			1/2 *		1/2	* Full dose as basal is also recommended
	HY short duration	1/3	1/3	1/3	Full			1/2 *		1/2	* Full dose as basal is also recommended
Wet land, direct seeded	General	1/3 *	1/3	1/3	Full			1/2 *		1/2	* For wet seeded, the first dose to be given 1 week after sowing. For dry seeded, first application to be given after establishment of the seedlings
	Mahsuri	1/3	1/3 *	1/3 **	Full			1/2		1/2 **	* 45 DAS, ** 85 DAS
Wet land, transplanted	HY, short duration	2/3		1/3 *	Full			1/2		1/2	* 5-7 days before PI
	HY, medium duration	1/2		1/2 *	Full			1/2		1/2 *	* 5-7 days before PI
	Mahsuri	1/2	1/4 *	1/4 **	Full			1/2		1/2 **	* 40 DAP, ** 60 DAP
Onattukara	General	1/2	1/4	1/4	Full			1/2		1/2	In very coarse soils, N & K may be given in five equal splits
Wyanad & hilly region	Long duration transplanted		1/2	1/2		Full *			1/2	1/2	* Along with first application of N & K
	Direct seeded		1/2 *	1/2		Full **			1/2*	1/2	* 45 days after seeding, ** With first application of N & K
Pokkali	General	Full*			Full*						* Apply entire quantity at the time of dismantling of mounds
Kootu-mundakan	First crop	1/2		1/2	Full			1/2		1/2	
	Second crop	Full*			Full*			Full*			*Entire quantity as single dose immediately after the harvest of first crop

BA = Basal application; AT = At tillering; API = At panicle initiation; HY = High yielding; DAP = Days after planting; DAS = Days after seeding

Baler: A **baler**, most often called a **hay baler** is a piece of farm machinery used to compress a cut and raked crop (such as hay, cotton, flax straw, salt marsh hay, or silage) into compact bales that are easy to handle, transport, and store. Often, bales are configured to dry and preserve some intrinsic (e.g. the nutritional) value of the plants bundled. Several different types of balers are commonly used, each producing a different type of bale - rectangular or cylindrical, of various sizes, bound with twine, strapping, netting, or wire.

- 20 kg and 30 kg pelletsbundles are tied up @ Rs 28/ bundle.

References

1. Rosamma, C.A., Elsy, C.R., Balachandran, P.V. and Rose, M.F., 2003. Pattambi rice varieties, Kerala Agricultural University.
2. FAO [Food and Agriculture Organization] Rome, 2011. http://faostat.fao.org/.
3. Thomas, J.J., 2010. Review of Agrarian Studies, Paddy cultivation in Kerala.

❏❏❏

Chapter 25

Pests of Major Crops

Gayathri P. Bose and **Hari Sankar S.S.** (Research Scholars)

Department of Agricultural Entomology, Kerala Agricultural University, College of Agriculture Vellayani, Thiruvananthapuram-695 522

1. Rice stem borer : *Scirpophaga incertulus* Pyralidae: Lepidoptera

The pest attacks all stages of the crop. The caterpillars bore into the stem and feed on the internal contents. Seedlings stage symptom appear as '**Dead heart**' and seed formation stage, it is **white ear head** (ear head with chaffy grains)

Adult moth

Dead heart and white ear

2. Rice bug *Leptocorisa acuta* Alydidae: Hemiptera

The adults and nymphs suck the milky contents of tender grains.

Individual rice grains become white and **chaffy**

Adult bug

Chaffy grains in panicle

3. Rice leaf roller *Cnaphalocrocis medinalis* Pyralidae: Lepidoptera

Caterpillars fold the rice leaves either longitudinally or transversely and scrape the green matter.

Adult moth and larvae within leaf roll

Leaf roll

4. Brown plant hopper *Nilaparvata lugens* Delphacidae :Hemiptera

Adults and nymphs of hoppers cluster together at the base of the plants just above water level and suck the sap, which results in yellowing and drying of plant. Dried plants are seen in circular patches in the field called as **'Hopper burn'**

Adult hoppers

Hopper burn

5. Rice Gall Midge *Orseolia oryzae* Cecidomyiidae: Diptera

Pinkish eggs laid at base of the leaf. Feeding by maggots result in circular leaf called **'silver shoot'or anecombu.**

From left to right: Adult fly, maggots within leaf sheath, silver shoot or anecombu symptom

6. Rice Case Worm *Paraponyx stagnalis* Pyralidae: Lepidoptera

The pale green caterpillars cut the leaf tips and fold them to form cases. They remain within the cases and feed.

Adult moth (left) and cut leaf cases floating on water (right)

Integrated pest management in rice

Adoption of an integrated programme having a mixture of eco-friendly tactics is the best option for pest management in rice fields.

- In pest endemic regions, grow tolerant varieties
- Monitor the pest population at weekly intervals, through light traps and insect net sweeps
- Keep the field and bunds free of weeds and grasses, as these act as alternate hosts
- Wider spacing and SRI method of cultivation reduce the attack of BPH
- Clipping tips of seedlings before transplanting to eliminate egg masses of rice stem borer
- Irrigation management: Alternate wetting and drying of the field reduce the population of BPH. Whereas case worms can be controlled by running a rope across and then drain away the stagnant water from the fields to flush out cases.
- Installation of pheromone traps or sleeve traps (Scirpo lure) @ 20/ ha 20 days after transplanting for the control of rice stem borer
- For the destruction of egg stages of rice stem borer or leaf roller, *Trichogramma japonicum* or *T. Chelonis* egg cards (Tricho cards) can be used at the rate of 2 cards/ acre 30 days after planting, replacing them at weekly intervals.

- Application of insecticides is recommended in heavily infested areas.
  - Nursery application of Fipronil 0.3% G @40 Kg/ha
  - Seedling root dip 0.02% Chlorpyriphos against rice gall midge
  - Acephate 75% SP - 1.6 g/l, Cartap hydrochloride 2g/l, Clorantraniliprole 18.5 % SC 3ml/10 l against stem borers and leaf rollers

Coconut

1. Rhinoceros Beetle *Oryctes rhinoceros* Scarabaeidae: Coleoptera

Adult beetle is large, black in colour with a pointed horn on its head. Eggs and grubs are found in manure pits. Adult feeds on central unopened spindle, which results in the formation characteristic **triangular cut** on fully opened fronts.

2. Red palm weevil *Rhynchophorus ferrugineus* Curculionidae: Coleoptera

This reddish brown weevil is the most destructive and damaging pest of coconut. Grubs bore into the trunk in large numbers, making tunnels. The symptoms appear as yellowing and wilting of leaves, **circular bore holes on stem with brownish ooze** and Foul odour from the crown. In the advanced stage of infestation, the **crown topples.**

Crown toppling

3. Coreid Bug *Paradasynus rostratus* Coreidae : Hemiptera

The bug generally clusters on leaf petiole, spathe or nuts and suck the sap. The attacked buttons become deformed with **crinkles and crevices on the husk and gummy exudations**.

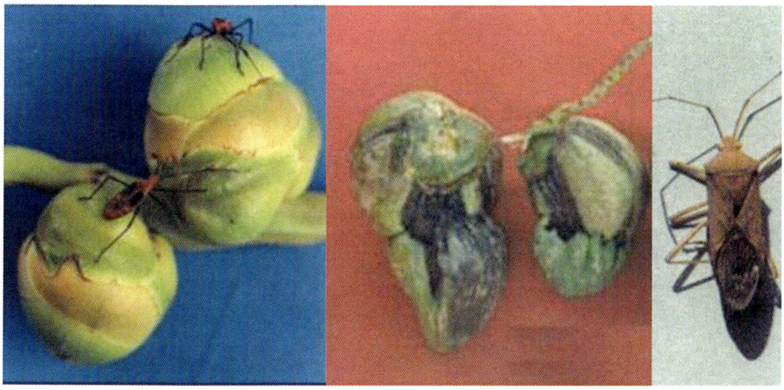

Malformed nuts and adults of nut crinkler

Coconut eriophyid mite *Aceria guerreronis* Eriophyidae: Acarina

These mites are invisible to naked eyes. They are seen in young buttons (2-6 month-old) under the perianth and suck sap. Initially, pale yellow **triangular patches** appear below the perianth. Later, the **patches turn brown** and become **necrotic lesions**. It reduces quality as well as quantity of nut.

Crinkled nuts with necrotic lesions

Integrated Pest Management in Coconut

- Periodic application of micro-nutrient mixture reduces the infestation of eriophyid mites.
- Hook out the beetles from palms with a beetle hook.

- Remove decaying plant materials from the plantation to prevent breeding of beetles.
- Harvest nuts regularly
- Avoid making steps or any other injury on the tree trunks to prevent egg laying by red palm weevil
- Use pheromone traps (Ferrolure or Sime RB) for attracting and killing rhinoceros beetle and red palm weevil @ 1 trap/2 ha.
- Fill the top most three leaf axils around the spindle with Sevidol 5G 25 g + fine sand 200 g or Naphthalene balls 10.5g (approx. 3 balls) and covered with fine sand or powdered neem cake once in 45 days.
- Inoculate the manure pits with entomopathogenic fungus, *Metarrhizium anisopliae* @ 2 Kg/ ton
- Place coconut log traps dipped in fermenting toddy or pineapple/sugarcane activated with yeast or molasses and insecticides like carbofuran in coconut plantations, to attract and trap red palm weevil.
- Infestation of sucking pests can be controlled by the application of systemic insecticides on coconut bunches.

* Integrated pest management should be applied as a community basis for a satisfactory results

Banana

Banana rhizome weevil *Cosmopolites sordidus* Curculionidae: Coleoptera

Black weevil, hide around banana plants during day time and are active during night. Eggs laid in small burrows in rhizome at ground level.

The emerging grubs bore into the **rhizome making extensive tunnels plugged with excreta.** Such rhizomes show black patches with rotting of tissues. Severe attack results in the **death of the unopened leaves** called the 'pipe'. **Undersized bunches and leaves** are another symptom of attack.

Tunnelling by grubs in rhizome

Death of unopened leaves

Pseudostem weevil *Odoiporus longicollis* Curculionidae: Coleoptera

Black weevil, that lay eggs in air chambers of leaf sheaths through slits cut on them. Grubs make extensive tunnels inside the pseudo stem and feedin on the internal tissue.

Gummy exudation from the oviposition punctures and from the feeding holes is the visible symptom. Yellowing and withering of leaves and undersized leaves are seen in the infested plants. Severely infested plants may **break and topple**.

Eggs laid in pseudostem

Gummy exudation and psudostem toppling

IPM in Banana

- Use healthy and pest free suckers for planting
- Avoid ratoon cropping
- Destroy the crop residues immediately after harvest of the crop
- Application of neem cake, daincha and cassia leaves to the pit at the time of planting
- Periodical removal of dried leaves from the plant to reduce hiding area of adult weevils

- Paring-Cut and remove the outer layer of rhizome and dry for 3-4 days after smearing it with slurry of cow dung and ash.
- Collect the weevils using pheromone trap (cosmo-lure)
- Place half meter long split banana psuedostem treated with *Beauveria bassiana* in the field during night. Collect and destroy the weevils in the morning.
- Soil application of *Beauveria bassiana* or *Metarhizium anisopliae* for the control of rhizome weevil.
- Leaf axil filling with 0.05% Chlorpiriphos or quinalphos to reduce the infestation of psuedostem weevil.
- Nanma and Menma (a tapioca leaf based organic pesticide) developed by CTCRI have been found to be very effective in managing psuedostem and rhizome weevil.

Vegetables

A. Borers

1. Shoot and fruit bores : *Earias vitella* (Bhindi), *Leucinodes orbonalis* (Brinjal), *Helicoverpa armigera* (Tomato)

The Caterpillar bore into the terminal shoots or fruits. The damaged shoots droop, wither and dry up. The infested **fruits are deformed with holes on them plugged with excreta.**

Fruit boring symptom in bhindi, brinjal and tomato

2. Pod borers *Maruca testulalis, Lampides boeticus, Helicoverpa armigera Etiella zincnella*

The larvae of pod borers, first scrape the surface of the pods and finally make holes in them and feed on the seeds. Bore holes on the pods are the peculiar symptom of the attack.

Pod borers in cowpea and their symptom of damage

3. Melon fruit fly *Bactrocera cucurbitae*

Adult fly lay eggs on the surafce of fruits and emerging maggots bore into the pulp of the fruit and feed within. The fruits become distorted and malformed initially. Later, the fruits rot and drop. A brown resinous fluid is seen oozing from the fruit. It is the most common and destructive pest of cucurbits.

Adult fly

Maggots

Infested fruits

B. Leaf feeders

1. Leaf eating caterpillars *Spodoptera litura, Diaphania indica, Anadevidia peponis*

Early instars of leaf eating caterpillars scrape off the green matter from the leaves and later instars defoliate the leaves.

Caterpillar damaging symptom

2. Leaf roller/ leaf webbers *Sylepta derogata*

Caterpillar rolls/webs the leaves the leaves together and they feed on the green matter by remaining with in the webs or rolls results in drying up of leaves.

Conical leaf rolls

Caterpillar with in leaf rolls

Leaf webs

C. Root feeders

1. Pumpkin beetle *Aulacophora* spp

Adults of pumpkin beetles feed extensively on the leaves. The grubs seen on soil, feed on the roots results in yellowing of leaves.

Adult beetle

Leaf symptom

D. Sucking pests

Sucking pests *viz.*, thrips, mealy bug, hoppers, white flies, aphids are found to cause serious damage to vegetables. Both, adults and nymphs suck the sap from the leaves and transmit viral diseases like leaf curl.

General feeding symptoms:

Thrips - upward curling of leaves

Aphids - Yellowing, crinkling and curling

White fly - Yellowing, sooty mould. It also transmits yellow vein mosaic disease in bhindi

Mites-White speckles on the leaves and in chilli, the symptom appears as downward curling of the leaves along with elongation of petiole (Rat tail)

Hoppers-Edges of infested leaves turn yellow and finally brown and crinkle severely, known as 'hopper burn'

Thrips

Aphid infestation on pods

White flies

Thrips attack

Hopper burn symptom

Rat tail symptom in chilli

IPM in Vegetables

- Select the crops according to the climatic conditions prevailing in the area
- Field application of lime 15 days before the date of planting
- Avoid continuous cropping with same variety or crop
- Remove alternate host plants and weeds
- Avoid excessive use of nitrogenous fertilizers
- Collect and destroy infested plant parts
- Use yellow sticky trap (especially for white flies and hoppers) @ 10/ha
- Pheromone trap, Cue lure trap @ 1 trap/15 cents controls cucurbits fruit fly and leucin lure helps to manage fruit and shoot borer in brinjal.

- Conserve natural enemies prevailing in the ecosystem. Spiders, lygaeid bug, Coccinellid beetles and syrphid maggots are naturally predaceous on leaf hoppers and aphids
- Three sprays of Neem oil – garlic emulsion 2% or nimbicidine or neemazal @ 1ml/l as a repellant before the start of infestation
- Spray entomopathogenic fungi *Lecanicillium lecanii* (and *Beauveria bassiana*
- Spray Bt formulation @ 2g/l for the control of caterpillar pests
- Spray Thiamethoxam 25 WG 2g/10 L or Imidacloprid 18.8 SL 3ml/l or Dimethoate 30 EC-1.25 ml/L or seed treatment with imidacloprid (200 SL 0.05% or 70 Ws 5g/kg) for the control of sucking pests.
- Flubendiamide 39.35 SC (1ml/10l) or Chlorantraniliprole 18.5 SC (3 ml/l) is effective against caterpillar pests
- Drenching of insecticides to the root zone is recommended for the control of root feeding insects

References

David, B.V. and Ramamurthy, V.V., 2012. Elements of Economic Entomology (7th edition). Namrutha Publications, Chennai, 390p.

Srivstava, K. and Dhaliwal, G. S., 2010. A Textbook of Applied Entomology Vol. II (3rd edition, Reprint), Kalyani Publishers, Ludhiana.

Atwal, A.S. and Dhaliwal, G.S., 2015. Agricultural pests of India and South-east Asia, Kalyani Publishers, Ludhiana.

❑❑❑

Chapter 26

Non-Insect Pests of Crops and Their Management

Hari Sankar S.S. and **Gayathri P. Bose** (Research Scholars)

Department of Agricultural Entomology, Kerala Agricultural University, College of Agriculture Vellayani, Thiruvananthapuram-695 522

Crop plants are subject to damage by diseases and various others animals like insects, birds, rodents and mites. Insect pests cause quantitative yield loss while rodents cause more of qualitative losses. The chapter is broadly divided into the following sections :

1. Mammalian pests
2. Avian pests
3. Mites
4. Snails, slugs and crabs

1. **Avian pests:** Some birds are closely associated with agricultural ecosystem for food and shelter. They usually damage by feeding of grains on field and drying yard, nibbling off ripe fruits, nesting in fruit trees, thus making harvesting difficult and products unhygienic. Some common avian pests are listed below:

Common name	Scientific name	Importance	Image
Baya weaver	*Ploceus philippinus*	Yellow birds that weaves nest using grass	
Bulbul	*Pycnonotus cafer*	Feeds on fruits,seeds, nectar and small insects. Sometimes kept as pets	
House crow	*Corvus splendans*	Closely associated with man, feeds on grains, fruits and dead animals. Vector of Newcastle disease	
Mynah	*Acridotheres tristis*	Feeds on berries, insects species	
Rose ringed parakeet	*Psittacula krameri*	Serious pests of sorghum, barley and cowpea. Raid field in large numbers and eat away the seeds wihin cowpea pods	

Blue rock pigeon	*Columba livia*	Scavengers, feed in flocks on human garbage. Vectors of avian influenza	
Tree pie	*Dendrocitta vagabunda*		

Management of bird pests

Avian pests can be managed only at a community basis. In places where they are serious menace, farmers should join hands to tackle them. Several management options are available. Choosing of any should be taken after considering their ecological and environmental impact, as many birds are integral part of many food chain. Some feasible management options are:

a) Bird scarers: These are objects that scare birds. They usually come in shapes of man, eagle, dead birds and brightly coloured balloons with scary eyes drawn on them.

b) Auditory repellants: Air horns, fire crackers, bird scarer guns are used. These usually produce explosive sounds that scares away birds in flock.

c) Visual repellants: Reflective tapes are tied around the borders of fields or tied as ribbons on stakes. They reflect sunlight and cause panic in birds, forcing them to fly away from the fields.

Different types of bird scarers

Reflective tapes for deterring birds

2. **Mammalian pests**: Mammals are animals that give birth to their young ones. Many mammals like cattle, pig, horses are associated with humans. Due to deforestation and destruction of natural habitats, certain wild mammals started frequenting agricultural fields. Common mammalian pests are:

Rats: They are usually called rodents. Their size varies from few centimeters in a **house mouse** to several centimeter long **bandicota**. They are nocturnal, feed on both, plant and animal matter and cause qualitative damage by excretion over heaped grains and other storage commodities, they also cut away ripe panicles of rice and take them to their burrows. Some rats even climb coconut trees and destroy nuts. They are a major pest in godowns. Due to their high fecundity and short intervals between reproductive cycles, they multiply exponentially in no time. They are vectors of bubonic plague.

Rat damage in coconut

Tanjore bow trap

Management

a) **Rat traps** : These are indigenously developed and modified from place to place like tanjore bow trap. There are two types of traps, one that kills instantly called back break or snap traps and another that traps only and requires killing later, called wonder trap. Multiple rats can be trapped in wonder trap. There are newer sticky traps that allow trapping mice at home.

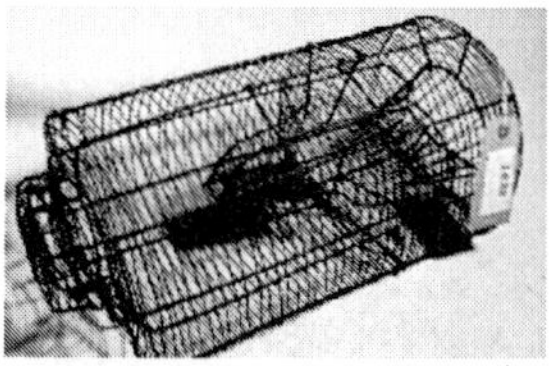

Wonder trap

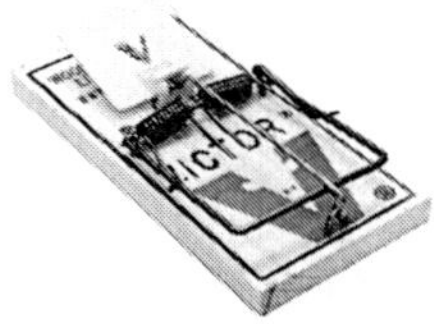

Snap/ back break trap

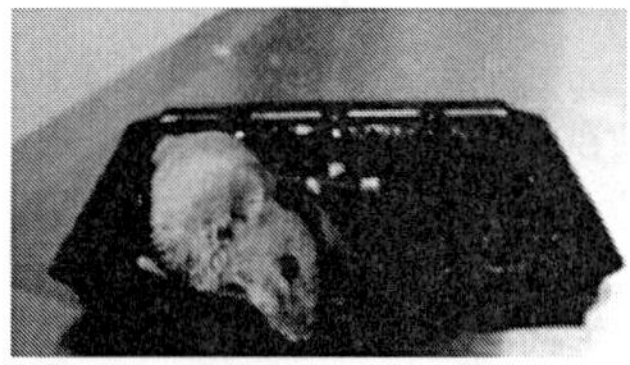

Sticky trap

b) **Bait poisons** : These are highly toxic chemicals mixed with attractive food and kept in places, where the rats are frequent. They offer clean and hassle free kill of rats. Two rat poisons are common in India: bromodiolone and zinc phosphide:

Wild hogs/ boar : They are serious pests of commercial crops near forest region. They attack during night and uproot tubers, eat away saplings and trample over crops ready to flower or harvest. These can be scared away by using fire cracker explosion, or electric fencing around the field.

Wild elephants : Elephants attack rice and sugarcane fields near to forests, particularly during flowering stage.

Wild hogs and elephants

*In case of pigs and elephants, it is advised to inform forest officials, prior to taking up any management options

3. **Snails and slugs** : These are usually aquatic organisms. They don't have a proper feet, they crawl over surfaces with a fleshy mass of muscle. The ones with shell are called as **snails** and those without outer shell are called **slugs**, but in large numbers, especially during rainy seasons. They usually feed on leaves and fruits, also cause damage due to the silvery trails left behind after crawling. **Giant African snail** was a serious menace few years back. They destroyes crops like banana, papaya, tuber crops and home gardens. They can be managed with metaldehyde, an attractant cum poison, specific to snails. They can also be trapped using wet gunny bags filled with vegetable waste.

Giant African snail and metaldehyde pellets

Crabs : Crabs in rice fields live in burrows near bunds. They cut young paddy plants at base and carry them to the burrows for feeding. They are nocturnal. They can be managed by use of bait traps made of fermented rice bran with any granular insecticide. Oilcakes can also be placed near the burrows:

4. **Mites:** They are usually microscopic, invisible to naked eyes, feed on lower surface of leaves results in crinkling and curling of leaves and buds, yellowing, bronzing, crumbling and malformation of affected parts. In some crops, they exhibit certain peculiar symptoms.

Sl. No	Crop	Symptoms	
1.	Coconut	Triangular yellow patch, brown discolouration 	
2.	Chilli	Downward curling of leaves with elongation of leaf petiole – '**Rat tail symptom**'	
3.	Jasmine	White hairy outgrowths on the leaf surface known as **'Erinium'**	
4.	Cinnamon gall mite	Irregular galls on underside of the leaves	
5.	Brinjal mite	White speckles on the upper surface of leaves	

They can be managed by the following methods:

- Nutrient management: Apply recommended dose of nitrogen, farmyard manure and micro-nutrients
- Spray entomopathogenic fungi talc formulation of *Lecanicilium lecanii* or *Hirsutella thompsoni* @ 20 g/L
- Spray neem oil-garlic emulsion 2% (ensure under surface of leaves is also equally coated)
- Spray insecticides *viz.*, Spiromesifen or spirotetramat @ 1.2 ml/ 1.0 L.

References

David, B.V. and Ramamurthy, V.V., 2012. Elements of Economic Entomology (7th edition). Namrutha Publications, Chennai, 390p.

Srivstava, K. and Dhaliwal, G.S., 2010. A Textbook of Applied Entomology Vol. II (3rd edition, Reprint), Kalyani Publishers, Ludhiana.

Atwal, A.S. and Dhaliwal, G.S., 2015. Agricultural pests of India and South-east Asia, Kalyani Publishers, Ludhiana.

❑❑❑

Chapter 27

Vermicomposting and Coirpith Composting for Income Generation

Kavya S.R. and **Greeshma Susan Mathew** (Research Scholars)

Kerala Agricultural University, College of Agriculture Vellayani, Thiruvananthapuram-695 522

Human livelihood opportunities are often closely linked to soil fertility conditions in dry tropical regions of India. Notwithstanding a common social background and possessing almost similar natural resources, farmers of a locality exhibit surprisingly drastic dissimilarity in their economic status. In such situations, more often than not, the factor determining the economic well-being of farmers is soil fertility. In particular, the biological component of the soil, represented by living organisms and dead organic matter, is the major factor limiting the fertility of dry land soils. Therefore, improving the biological fertility is usually a priority of land development programmes. If activities designed for such purposes also afford an opportunity to earn a livelihood, the eventual benefit realised is much larger.

Composting is the process by which biomass is broken down to humus, which has several beneficial effects on soil physical and chemical properties. Further decomposition of humus releases the nutrients contained in it for crop uptake. A requirement for composting is organisms that feed on biomass and break it down to physically finer particles and chemically less complex substances. In nature, earthworms and micro-organisms decompose dead biomass. When this process is initiated with the deliberate introduction of earthworms into a stack of biomass, it is called vermicomposting.

Of the limited biomass available, farmers use straw, leaf litter and tender stem as fodder for farm animals while hardy stem and coarse leaf are sources of domestic fuel. The leftover biomass, after these immediate priorities are met, is coarse material that does not break down easily. In such material, earthworms are more effective in initiating the decomposition process and paving the way for subsequent microbial action.

Another source of biomass available to farmers is cattle dung. It is an excellent substrate that undergoes composting on its own and becomes farmyard manure of very high quality. Unfortunately, the dung available with small farmers does not compost well because of the state, in which it is available. Animals owned by them usually graze in the open and the dung is relatively dry, when collected. It is then left in a collection pit, where it loses more moisture. Therefore, dung collected in the open, unlike fresh dung, does not have the microbial populations in required numbers to decompose into farmyard manure. However, it becomes an excellent substrate for vermi-composting, when wetted.

Earthworms are liberated and cost effective farm relief. The worms are accountable for a variety of elements including turning common soil into superior quality. They break down organic matter and when they eat, they leave behind castings that are an exceptionally valuable type of fertilizer. The earthworm is one of the nature's pinnacle "soil scientists". *Eudrillus eugineae* has been identified as the most ideal species of earthworm for vermitechnology under Kerala conditions.

Mechanism of vermicomposting

Materials consumed by worms undergo physical breakdown in the gizzard resulting in particles <2 μ, giving thereby an enhanced surface area for microbial processing. This finally ground material is exposed to various enzymes, such as protease, lipase, amylase, cellulase and chitinase secreted into lumen by the gut wall and associated microbes. These enzymes breakdown complex bio-molecules into simple compounds. Only 5-10% of the ingested material is absorbed into the tissues of worms for their growth and rest is excreted as cast. Mucus secretions of gut wall add to the structural stability of vermicompost.

Procedure

It is mostly prepared in either pit or heap method. The dimensions of either heap or pit are 2.5 x 1 x 0.3 m. They are taken in thatched sheds with sides left open. The length and width can be increased or decreased depending on the availability of material but not the depth because the earthworm's activity is confined to limited depth only.

First of all, select a site which is not under any economic use and is shady with no water stagnation. The site should be near to a water source. The bottom and sides of the pit are made hard by compacting with a wooden mallet. At the bottom of the pit, a layer of coconut husk is spread with the concave side upward to ensure drainage of excess water and for proper aeration The husk is moistened and above this, bio-waste mixed with cowdung in the ratio of 8:1 is spread up to a height of 30 cm above the ground level and water is sprinkled daily. After the partial decomposition of wastes for 10 to 14 days, the worms are introduced @ 500 to 1000 numbers per pit. The pit is covered with coconut fronds. Moisture is maintained at 40 to 50 per cent. When the compost is ready, it is removed from the pit along with the worms and heaped in shade with ample light. The worms will move to bottom of the heap. After one or two days, the compost from the top of the heap is removed. Put back the undecomposed residues and worms to the pit for further composting, as described above. The vermicompost produced has an average nutrient status of 1.5% N, O.4% P_2O_5 and 1.8% K_2O with pH ranging from 7.0 to 8.0. The nutrient level will vary with the type of material used for composting.

Precautions

1. The compositing area should be provided with sufficient shade to protect from direct sunlight.
2. Adequate moisture level should be maintained by sprinkling water, whenever necessary.
3. Take preventive measures to ward off predatory birds, ants or rats.

Normal time taken for vermicomposting is 60-75 days. When the compost is ready, watering should be stopped to facilitate separation of worms from the compost. The compost is taken out from the pit and heaped in sunlight. The worms settled at the bottom of the heap are separated and used for further composting. It is better to apply the fresh compost directly to the crop or can be dried under shade and sieved for storage.

Compost in progress

Eudillus eugine- worms used in compost

Vermicomposting of household wastes

Select a wooden box of 45cm x 30cm x 45 cm or an earthen/ plastic container with broad base and drainage holes. Keep a wire net with small holes at the bottom of the box. Add a layer of soil of 3 cm depth and a layer of coconut fibre of 5 cm depth above it for draining of excess moisture. About 250 worms are introduced into the vermicomposting unit alongwith a handful of cow dung. Spread daily vegetable wastes in layers. Cover the top of the box with a piece of jute sac to provide dimlight inside the box. When the box is full, keep the box without disturbance for a week. When the compost is ready, keep the box outside for 2-3 hours so that the worms come down to the lower fibre layer. Remove compost from the top, dry and sieve. The vermicompost produced has an average nutrient status of 1.8% N, 1.9% P_2O_5 and 1.6% K_2O, but composition will vary with the substrate used.

Earthworms can be multiplied in 1:1 mixture of cowdung and bio-waste taken in a cement tank or wooden box or plastic basin with proper drainage facilities.

Vermicompost in bins

Compost after seiving- ready for a use

Coirpith composting

Coirpith, one of the agricultural wastes is produced and heaped in large quantities as waste material from the coir industry. Approximately 2.5 lakh tonnes of coirpith accumulate in Kerala as waste. Coirpith has wide C: N ratio and its lignin rich nature does not permit natural composting process as in other agricultural wastes. *Pleurotus* spp. has the capacity to degrade part of the cellulose and lignin present in coirpith by production of enzymes, viz., celluloses and lactases. The C: N ratio of coirpith is reduced from 112:1 to 24:1 as a result of composting. The lignin content also reduces from 30% to 4%.

Method of composting

Materials required: Corpith 1 tonne, poultry manure 10 kg, mushroom (*Pleurotus*) spawn 1.5 kg.

Select a shaded place of 5 x 3 m dimension and level it after removing weeds. First spread 100 kg coirpith uniformly. Spread 300g (one bottle or cover) of *Pleurotus* spawn on this and cover this with a second layer of 100kg coirpith. On the surface of the second layer, spread 2kg poultry manure uniformly. Repeat this sandwiching process of one layer of coirpith with spawn, followed by another layer of coirpith with poultry manure upto 1 m height. Sprinkle water, if necessary to keep the heap moist. Allow the heap to decompose for two months. This coirpith compost contains macro-nutrients as well as micro-nutrients. It has the unique property of absorbing and retaining moisture to about 500-600 per cent. It improves the water infiltration rate and hydraulic conductivity of soil.

Coir pith composting

Livelihood opportunity

Vermicomposting is picking up as an enterprise, both at the individual as well as the group level. At an individual level, farmers are not only benefitting by the improvement of soil fertility through its application on land but also they are producing more than required for their land and selling surplus to other farmers. For example, a progressive farmer of Khushal Gari village, Atrauli Block who owns a nursery of horticulture plants earned an income of around US$ 2500 through production and sale of earthworms and vermicompost, besides meeting his own nursery requirements.

As a group initiative, vermicomposting is being taken up at different levels. For example, in Barola Hagi village of Lodha block (about 20 kms away from the city), a co-operative society of SC/ST people having cattle, is managing a vermiculture unit with around more than hundred vermicompsting beds. Most of the vermicompost produced is being sold, as they do not have much agricultural land for field application. Similarly, some of the enterprising graduate students have established their own units, enabling them to meet their contingency requirements. Realizing the success of this programme, rural banks of the district have come forward to associate themselves for the technology dissemination, encouraging more and more farmers to adopt vermicomposting.

Self-help group approach is developing in a major way in vermicomposting. Village banking institutions are highly impressed with the micro-credit system established by the village communities. Good linkages have been developed in the village cluster for marketing the worms and compost produced through self-help groups. Trained beneficiaries are being financed for the establishment of vermicompost units and vermiculture to propagate the activity in the village at large scale.

Mass rearing and maintaining worm cultures and tapping of organic wastes for their maintenance has a good scope for developing it as a cottage industry in developing country like India. The effort is also to mitigate the ill effects of farming intensification and use of inorganic fertilizers and pesticides.

❑❑❑

Chapter 28

Integrated Nutrient Management for Sustaining Soil Health and Quality

Dr. B. Aparna and **Dr. Gladis, R.** (Asst. Professor)

Department of Soil Science & Agri. Chemistry, Kerala Agricultural University, College of Agriculture Vellayani, Thiruvananthapuram-695 522

Soil is the key source of food production and soil resources play a key role in determining the sustainable productivity of an ecosystem. Indiscriminate use of high analysis chemical fertilizers has resulted in the deficiency of nutrients and decline in organic carbon. The continuous use of chemical fertilizers over a long period may cause imbalance in the microflora and therefore, indirectly affect the biological properties, such as enzyme status ultimately, it may have an adverse effect on the soil leading to soil degradation. The continuous application of chemical fertilizers without organic matter will cause a deteoriation in soil properties and reduction in productivity of the soil. However, when organic manures are applied in conjunction with chemical fertilizers for growth of crop, decline in organic carbon status was arrested and the gap between the potential yield and actual yield will be bridged to a larger extent.

Massive use of fertilizers and pesticides supported by heavy subsidies as a part of "Green Revolution Strategy" has led to the present problems. The health of mother earth is also deteriorating day-by-day. The problems and toxicities can be tackled by adopting alternate technologies and management of soil without any harm to the ecology such as Organic Farming, Integrated Nutrient Management (INM), Integrated Pest and Disease Management (IPDM), etc. Presently alternative farming strategies viz. organic farming, low input farming is being increasingly studied to sustain the soil resource base even while meeting the needs and concerns.

Sustainable soil nutrient-enhancing strategies involve the wise use and management of inorganic and organic nutrient sources in ecologically sound production systems. The primary goal of Integrated Nutrient Management (INM) is to combine old and new methods of nutrient management into ecologically sound and economically viable farming systems that utilize available organic and inorganic sources of nutrients in a judicious and efficient way. Integrated nutrient management optimizes all aspects of nutrient cycling. It attempts to achieve nutrient cycling with synchrony between nutrient demand by the crop and nutrient release in the soil, while minimizing losses through leaching, runoff, volatilization and immobilization. Integrated Nutrient Management refers to the maintenance of soil fertility and of plant nutrient supply at an optimum level for sustaining the desired productivity through optimization of the benefits from all possible sources of organic, inorganic and biological components in an integrated manner.

The term "IPNM" is interpreted in the much broader more holistic sense of "land husbandry". It, thus, embraces soil, nutrient, water, crop, and vegetation management practices, tailored to a particular cropping and farming system, undertaken with the aim of improving and sustaining soil fertility and land productivity and reducing environmental degradation. Integrated Plant Nutrient Management aims to optimize the condition of the soil, with regard to its physical, chemical, biological and hydrological properties, for the purpose of enhancing farm productivity, whilst minimizing land degradation.

Objectives of INM

The major objective of this system is the integral use of organic sources of nutrients, such as vermicompost, bio-fertilizers with chemical fertilizers for optimizing yield. Integrated use of organic manures and chemical fertilizers has been found to be promising in arresting the decline in productivity under modern intensive farming through the correction of marginal deficiencies of certain secondary and micro-nutrient elements in the course of decomposition of organic manures on one hand and its beneficial influence on physical and biological properties of soil on the other hand. Hence, Integrated Nutrient Management aims at maintenance of soil fertility for sustained crop productivity through rational and complementary use of organic, inorganic and biotic sources of plant nutrients required for crop growth and quality in an integrated manner appropriate to each cropping system and farming situation in its ecological, social and economic viabilities.

Need for Integrated Nutrient Management

Farming these days need drastic changes owing to change in policies and human needs. With the advent of green revolution, a serious threat to the environment has been witnessed. Fuelled by ever increasing population to produce more food and fiber, to meet the demands from the limited rather shrinking land area, man's activity has resulted in degradation beyond restoration. Signs of stress on ecosystem are everywhere today reported in the press. Understanding the underlying biological process in tandem with identification of indicators of the above ecosystem stress is inevitable to promote long term sustainability of the ecosystem.

Integrated Nutrient Management (INM) is an alternate viable strategy targetting on sustainable production and the application of organic matter is often claim effective method of maintaining soil fertility. The basic principle behind INM concept is to supply both, chemical and organic fertilizer for sustainable crop production. The objectives of this approach are efficient and judicious use of all major sources of plant nutrients so as to get maximum economic yield, by maintaining stability in crop production through the correction of marginal deficiencies of secondary and micro-nutrients on one hand and maintaining soil health on the other hand.

Balanced nutrition involves systematic exploitation of potential of soil resources, chemical fertilizers, bio-fertilizers and organic manures. When manure was applied in conjunction with chemical fertilizers for efficient growth of crop, decline in organic carbon was arrested and the gap between potential yield and actual yield was bridged to a large extent. The addition of organic manures with chemical fertilizers may simulate mineralization and immobilization of plant nutrients thereby affecting their amounts in different organic and inorganic forms in soil.

Components of Integrated Nutrient Management

Fertilisers

Fertilizers are considered to be the most important component of INM. Judicious use (i.e. lower rates, split application, banding) of inorganic fertilizers is needed on infertile soils, to sustain high crop yields and maintain an optimum balance of nutrients. Continuous use of relatively high rates of nitrogen fertilizers on can reduce soil pH (acidification) and seriously reduce soil fertility. Acidification occurs mainly through the loss of exchangeable bases in leaching (Ca, Mg, K) and acid production during aluminium ion hydrolysis and nitrification. The dependence on fertilizers has been increasing constantly because of the need to supply large amount of nutrients in intensive cropping with high productivity. Nutrients exported from the soil through harvested biomass or lost from soil by gaseous loss, leaching, or erosion must be replaced with nutrients from external sources. The judicious use of chemical fertilizer is essential to maintain soil fertility. Nonetheless, fertiliser consumption is not only inadequate but also imbalanced. While organics and biofertilisers are expected to bridge a part of this gap, efficient use of fertilisers in narrowing the nutrient supply gap also needs greater emphasis. Utilization of fertiliser nutrients by the crops vary from 30-50% in case of N, 15-20% in case of P and less than 5% in case of micro-nutrients. Thus, substantial amount of applied nutrients is lost through various pathways.

Organic Manures

Organic manures like urban compost, FYM, crop residues, human excreta, city refuse, rural compost, sewage-sludge, pressmud and other agro industrial wastes have large nutrient potential. Compost and FYM have traditionally been the important manures for maintaining soil fertility and ensuring yield stability. Other potential organic sources of nutrients, such as non-edible oilcakes and wastes from food processing industry are also there. Moreover, there are several industrial by-products and municipal wastes with fair nutrient potential.

Bio-fertilizers

Bio-fertilizers are defined as preparations containing living cells or latent cells of efficient strains of micro-organisms that help crop plants' uptake of nutrients by their interactions in the rhizosphere, when applied through seed or soil. They accelerate certain microbial processes in the soil which augment the extent of availability of nutrients in a form easily assimilated by plants. Bio-fertilizers, more commonly known as microbial inoculants, are artificially multiplied cultures of certain soil organisms that can improve soil fertility and crop productivity. Although the beneficial effects of legumes in improving soil fertility was known since ancient times and their role in biological nitrogen fixation was discovered more than a century ago, commercial exploitation of such biological processes is of recent interest and practice. The commercial history of bio-fertilizers began with the launch of 'Nitragin' by Nobbe and Hiltner, a laboratory culture of Rhizobia in 1895, followed by the discovery of Azotobacter and then the blue green algae and a host of other micro-organisms. Azospirillum and Vesicular-Arbuscular Mycorrhizae (VAM) are fairly recent discoveries.

Rhizobium (RHZ): These inoculants are known for their ability to fix atmospheric nitrogen in symbiotic association with plants forming nodules in roots (stem nodules in Sesabania rostrata).RHZ are, however, limited by their specificity and only certain legumes are benefitted from this symbiosis.

Azotobacter (AZT): This has been found beneficial to a wide array of crops covering cereals,millets, vegetables, cotton and sugarcane. It is a free living and non-symbiotic nitrogen fixing organism that also produces certain substances good for the growth of plants and antibodies that suppress many root pathogens.

Azospirillum (AZS): This is also a nitrogen-fixing micro-organism beneficial for non-leguminous plants. Like AZT, the benefits transcend nitrogen enrichment through production of growth promoting substances.

Blue Green Algae (BGA) and Azolla: BGA are photosynthetic nitrogen fixers and are free living. They are found in abundance in India. They too add growth-promoting substances including vitamin B_{12}, improve the soil's aeration and water holding capacity and add to biomass, when decomposed after lifecycle. Azolla is an aquatic fern found in small and shallow water bodies and in rice fields. It has symbiotic relation with BGA and can help rice or other crops through dual cropping or green manuring of soil.

Phosphate solublizing (PSB) /Mobilizing bio-fertilizer: Phosphorus, both native in soil and applied in inorganic fertilizers becomes mostly unavailable to crops because of its low levels of mobility and solubility and its tendency to become fixed in soil. The PSB are life forms that can help in improving phosphate uptake of plants in different ways. The PSB also has the potential to make utilization of India's abundant deposits of rock phosphates possible, much of which is not enriched.

Vesicular Arbuscular Mycorhiza (VAM)

VAM improves the uptake of P by inoculated plants.There is enhanced absorbtion of water and other nutrients, such as N, K and certain other micronutrients.

Plant Growth Promoting Rhizobacteria (PGPR)

The group of bacteria that colonize roots or rhizosphere soil and beneficial to crops are referred to as plant growth promoting rhizobacteria (PGPR).The PGPR inoculants currently commercialized that seem to promote growth through at least one mechanism; suppression of plant disease (termed Bio-protectants), improved nutrient acquisition (termed Bio-fertilizers), or phyto hormone production (termed Bio-stimulants). These PGPR are referred to as Bio-stimulants and the Phytohormones they produce include indole-acetic acid, cytokinins, gibberellins and inhibitors of ethylene production.

Management of Crop Residues

Organic nutrient sources include plant residues, leguminous cover crops, mulches, green manure, animal manure, and household wastes. Under continuous cropping, recycling and reusing nutrients from organic sources may not be sufficient to sustain crop yields.

Strategies and practices for soil organic matter management include:

- Returning organic materials to the soil, to replenish soil organic carbon lost through decomposition (recycling of plant and animal residues, green manuring, cover crop rotation).
- Ensuring minimum disturbance of the soil surface (residue mulch, conservation tillage) to reduce the rate of decomposition.
- Reducing soil temperature and water evaporation by mulching the soil surface with plant residues; and
- Integration of multi-purpose trees and perennials into cropping systems to increase the production of organic materials.

Green Manure Crops and Inter-cropping

Timely applications of organic materials with a low C/N ratio, such as green manure and compost, could synchronize nutrient release with plant demand and minimize the amount of inorganic fertilizer needed to sustain high crop yields for short-cycle crops, such as maize, rice, and soyabean, all of which have a high nutrient demand. Fast-growing leguminous species, such as Mucuna (Mucuna utilis) and Kudzu (Pueroria phseoloides) can be especially useful as cover crops for erosion control, weed suppression and for soil fertility restoration. Leguminous green manures and cover crops are able to:

- Enrich the soil with biologically fixed N;
- Conserve and recycle soil mineral nutrients;
- Provide ground cover to minimize soil erosion, and
- Require little or no cash input.

Other classical examples are:Sun hemp (Crotalaria juncea), Daincha (Sesbania aculeate and Sesbania rostrata), Wild indigo or Kolingi (Tephrosia pur purea), Gliricidia (Gliricidia maculata), Subabul (Leucaena leucocephala), Cassia (Cassia auriculata).

Agroforestry

Agroforestry refers to all forms of land-use systems in which trees or woody perennials are deliberately planted on the same land management unit in association with livestock and/or annual crops, with significant ecological interactions between the woody and non-woody components. Agroforestry systems, in which tree legumes are inter-planted with cereals can be effective in soil nutrient cycling and enhancement. The benefits of tree-annual crop association arc:

- Retrieval of nutrients from below the rooting zone of annual crops.
- Reduction of nutrient losses from leaching, runoff and erosion.
- Legume trees increase the supply of nutrients within the rooting zone of annual crops through N input by biological N_2 fixation.

Conclusion

Integrated Nutrient Management (INM) is, thus, a recent strategy for enhancing the availability of applied as well as native soil nutrients, synchronizing the nutrient demand of the crop with nutrient supply from native and applied sources, providing balanced nutrition to crops and minimizing the antagonistic effects resulting from hidden deficiencies and nutrient imbalance, minimizing the deterioration of soil, water and ecosystem by promoting carbon sequestration, reducing nutrient losses to ground and surface water bodies and to atmosphere thereby improving the soil physical, chemical and biological quality parameters leading to the sustainability of soil health.

References

1. Jackson, M.L., 1952. Soil Chemical Analysis. Prentice Hall of India Pvt. Ltd. New Delhi, p. 111-203.
2. Lakshminarayana, K. and Patiram, K., 2005. Influence of inorganic, biological and organic manures on yield and nutrient uptake of groundnut and soil properties. *Indian J. Agric. Sci.*, **75**(4): 218-221.
3. Dwivedi, B.S., Singh, V.K. and Dwivedi, V., 2004. Application of phosphate rock, with or without Aspergillus awamori inoculation, to meet P demands of rice-wheat systems in the Indo-Gangetic plains of India. *Aus. J.Exp.* Ag. 44: 1041-1050.
4. Roy, R.N. and Ange, A.L., 1991. In Integrated Plant Nutrition System (IPNS) and Sustainable Agriculture, pp. SV/1-1-SV/1-12, *FAI,* New Delhi.
5. Singh, M., Dwivedi, B.S. and Datta, S.P., 2012. Integrated nutrient management for enhancing productivity, nutrient use efficiency and environmental quality. *In: Soil Science in the Service of Nation,* pp. 55-67, ISSS, New Delhi.
6. Yadav, R.L., Singh, S.R., Prasad, K., Dwivedi, B.S., Batta, R.K., Singh, A.K., Patil, N.G. and Chaudhary, S.K., 2000. In Yadav JSP and Singh GB (eds), *Natural Resource Management for Agricultural Production in India,* pp. 775-870, ICMNR, New Delhi.

□□□

Chapter 29

Weed Management

Dr. Ameena M. (Asst. Professor)

Department of Agronomy, Kerala Agricultural University, College of Agriculture Vellayani, Thiruvananthapuram-695 522, Kerala

Weed management is the application of certain principles and suitable methods that will improve the vigour and uniform stand of the crop. At the same time, ignore or discourage the invasion and growth of weeds.

Methods of Weeds Management

Weed control and weed management are the two terms used in weed science. Weed control is the process of limiting infestation of the weed plant so that crops can be grown profitably, where as weed management includes prevention, eradication and control by regulated use, restricting invasion, suppression of growth, prevention of seed production and complete destruction. Thus, weed control is one of the aspects of weed management.

Principles of Weed Management are:

1. Prevention
2. Eradication
3. Control

Prevention

Prevent the entry and establishment of weeds into uninfected area.

Eradication

It is complete removal of all live plant parts and seeds of the weed from an area. It may be a field/farm/village/geographical region depending upon the need.

Control

Weed infestations are reduced but not necessarily eliminated. Weed control includes :

1. Physical
2. Cultural
3. Biological
4. Chemical

Prevention

It encompasses all measures taken to prevent the introduction and/or establishment and spread of weeds. Such areas may be local, regional or national in size. No weed control programme is successful, if adequate preventive measures are not taken to reduce weed infestation. It is a long, term planning so that the weeds could be controlled or managed more effectively and economically than is possible, where these are allowed to disperse freely.

Arresting the movement and denying the entry in uninfested area. Crop hygiene can also be considered as prevention. Weeds producing seeds in current season are the inoculum for next season. If some how they do not produce seeds in current season, this is also one of the preventive measures. Thus, controlling weeds at flowering stage, which will prevent contamination in other season, is also a preventive measure.

Important weed prevention practices are:

1. Use clean crop seeds/weed free crop seed
2. Avoid feeding of straw, grain or hey containing weed seeds to livestock without destroying their viability by grinding, cooking and ensiling.
3. Use well-rotten\decomposed organic manure. Avoid reaching of weed seeds into the compost pit
4. Prevent movement of weeds with other farm resources
5. Keep non-crop area clean
6. Keep vigilance
7. Follow legal & quarantine measures

Use clean crop seeds / weed free crop seed

The prevention of weeds that disperse with crop seed can be achieved in two ways by the production of weed free crop seed at govt. farms or at farmer's fields and cleaning of the crop seeds before storage and at the time of sowing In seed production plots, weeds are removed before flowering.

Physical and mechanical methods

These methods include under non-chemical method of weed control. Implements used vary from simple to multiple tractor drawn implements.

The mechanical methods include:

1. Hand weeding
2. Hand hoeing
3. Puddling
4. Digging
5. Sickling
6. Dredging and chaining
7. Flooding
8. Burning
9. Soil Solarisation
10. Cheeling
11. Tillage
12. Mulching

Hand weeding

Removal of weeds either manually or by using tools like khurpi or sickle, when weeds grown upto some extent. Effective against annuals and biennials and controls only upper portion of the perennial. Higher labour is required and is tiresome.

Digging

Digging is useful for patch or spot control of obnoxious / perennial weeds. Digging is very useful in the case of perennial weeds to remove the underground propagating parts of weeds from the deeper layer of the soil. They can be eliminated by digging with crowbar or pick axe, etc. For large areas, it is not desirable because it is costly and labour-oriented.

Mowing

It is cutting of uniform growth from the entire area upto the ground level. It is useful more in non-cropped areas than cropped areas. Mowing improves aesthetic value of an area. Effective against erect and herbaceous weeds.

Cutting

Cutting is the topping/cutting of the weeds little above ground level. It is done with help of axes and saws. It is mostly practiced against brushes and trees. In aquatics under water, weed cutters are used.

Dredging

This is used to control aquatic weeds growing in shallow ditches. Mechanical pulling of aquatic weeds alongwith their roots & rhizomes from the mud.

Chaining

Very big & heavy chain is pulled over the bottom of a ditch with tractors alongwith embankments of ditch. With rubbing action of chain, weeds can be fragmented & collected by nets and hooks.

Burning

It is cheapest method to eliminate the mature unwanted vegetation in non-cropped areas and range lands. Coagulation of protoplasm occurs with which plant dies.

Flaming

It is the momentary exposure of green weeds to as high as 1000°C from flame torches to control in row weeds. e.g. Flaming is used in western countries for selective weed control in crops like cotton, onion, soybean and fruit orchards.

Soil Solarization

It is also called solar soil heating. It is effective against weeds which are produced from seeds. It doesn't involve any tillage of the field. Covering the soil with transparent, very thin plastic sheets of 300 micron thickness during hottest part of summer months for 2-4 weeks.

This increases the temperature by 10-12°C over the unfilmed control fields. Then weeds seeds are desiccated, which are present at top 5 cm soil depth.

Tillage

Tillage is done for preparing good seedbed, conservation of soil moisture & weed control. Tillage removes weeds from the soil resulting in their death. It may weaken plants through injury of root and stem pruning, reducing their competitiveness or regenerative capacity: Pre plant tillage helps in burying the existing weeds. Bring the seeds to the soil surface for germination and their subsequent destruction by suitable secondary tillage implements. Incorporation of pre-plant herbicides. Post-plant tillage (row cultivation) helps in mixing of manures and fertilizers & control of weeds, soil and water conservation.

Mulching

Principle is exclusion of sunlight from environment. Polythene Sheets, natural materials like paddy husk, ground nut shells, saw dust, etc. are used as mulching material. The thickness should be enough to cut off light (i.e. 10-15 cm). The efficiency of polythene sheet is more (more polythene), if it is applied in continuous sheet rather than in splits. It is effective against annual weeds and perennial weeds like *cynodon dactylon* and *sorghum halopense.*

Cultural Practices / Crop Husbandary Practices

Cultural methods, alone cannot control weeds, but help in reducing weed population. They should, therefore, be used in combination with other methods:

1. Proper crop stand and early seedling vigour
2. Selective crop simulation
3. Proper planting method
4. Planting time
5. Crop rotation
6. Stale seedbed
7. Smother crops
8. Minimum tillage

9. Summer fallowing
10. Lowering area under bunds
11. Flooding and drainage

Crop rotation

Growing of different crops in recurrent succession on the same land is called as crop rotation. Monocropping favours persistence and association of some weeds. Crop rotation is effective in controlling of crop associated and crop bound weeds, such as *Avena fatua* in wheat and *Cuscuta* in dodder. Wheat-pea and gram break the *Avena* in wheat, Lucern - grain crop rotation control *Cuscuta.* The obnoxious weeds like *Cyperus rotundus* can be controlled effectively by including low land rice in crop rotation.

Stale Seedbed

It is the one where one or two flushes of weeds are destroyed before planting of any crop. This is achieved by soaking a well-prepared field with either irrigation or rain and allowing the weeds to germinate. These weeds are controlled by using herbicides like glychosate and by mechanical methods, then sow the crop. Here the advantage is the crop is germinated in weed free environment. In this way, weed seed bank is exhausted.

References

Weeds Rao,V.S., 2000. Principles of weed Science. Oxford and IBH Publishing Company, New Delhi.

Thomas, C.G and Abraham, C.T., 1998. Common weeds of Rice ecosystem and their management. KAU, Thrissur FC.

De Datta, S.K., 1981. Principles and Practices of Rice Production. John Wiley and Sons, New York.

Singh, C., Singh, P. and Singh, R., 2003. Modern Techniques of Raising Field Crops. Oxford and IBH, New Delhi.

□□□

IV. Crops Improvement

Chapter 30

Plant Tissue Culture

Dr. Swapna Alex (Professor & Head) and **Dr. K. B. Soni** (Professor)

Department of Plant Biotechnology, Kerala Agricultural University, College of Agriculture Vellayani, Thiruvananthapuram-695 522

Plant Tissue Culture is an important aspect of Plant Biotechnology that involves techniques for large scale production of planting materials and crop improvement. It is based on the concept of **totipotency,** which states that a plant cell has the potential ability to develop into a whole organism, if provided with suitable conditions. The Austrian botanist Gottlieb Haberlandt is considered as the Father of Plant Tissue Culture.

In plant tissue culture, explants like plant cell, tissue or organ are cultured under aseptic conditions in a defined or semi-defined medium. The cultures are maintained in solid (with agar as solidifying agent) or liquid medium. Medium contains all the major, minor and micro elements including vitamins for the continued growth of the explants. Carbohydrate like sucrose is provided as major energy source. The growth of the explants in the *in vitro* condition is regulated by plant growth regulators like cytokinin, auxin and gibberrellins. The cultures are incubated under controlled conditions of temperature, light and relative humidity.

Types of cultures: The main types of plant cultures are callus culture, embryo culture, single cell culture, organ culture, protoplast culture and suspension culture.

Micro-propagation: One of the main applications of Plant Tissue Culture is large scale propagation. It is mainly done through micro-propagation. This technique provides large number of genetically uniform plants from micro-sized explants usually apical meristem in a very short period of time. It is one of the main methods being currently adopted in many crop plants for quality planting material production.

Stages of micro-propagation: Micro-propagation technique consists of different stages as follows:

Stage I Maintenance of the stock plant: We select an elite mother plant and maintain it as a source of explants.

Stage II Culture establishment: The explants are collected from the mother plant, washed and surface sterilized with chemicals and inoculated into medium.

Stage III: Multiplication of progagules: Depending on the type of explants and its response different routes of multiplication (axillary bud proliferation, somatic organogenesis or somatic embryogenesis) can be followed. This stage includes many rounds of subculturing to obtain sufficient number of shoots.

Stage IV Rooting: The shoots developed are transferred to root inducing medium or rooted *ex vitro*.

Stage V: Hardening: The rooted plantlets are gradually acclimatized to the field condition, during which the physiological abnormalities are corrected.

Culture Medium: The cultures are grown under *in vitro* conditions in an artificial medium which can be defined or semi-defined. In a defined medium, all the components will be known and the results are usually reproducible. Whereas, in a semi-defined medium natural extracts, such as coconut milk, fruit juices, etc., the composition of which are not exactly known, may also be added in the medium for better results. However, the reproducibility of the results may be affected. Different types of media have been developed, such as B5 medium, MS medium, Whites medium and Woody Plant medium. Among them MS medium developed by the scientists Toshio Murashige and Folke K. Skoog in 1962 is commonly used for culturing of different crop plants with slight modifications.

The components of the tissue culture medium include salts of major and minor nutrients, carbon source, vitamins, amino acids and hormones. The most commonly used carbon source is sucrose and the essential vitamin is Thiamine. Nitrogen is usually provided in three forms viz., ammoniacal, nitrate and organic forms to ensure the availability of nitrogen throughout the growth of the culture. The *in vitro* response of the plant mainly depends on the hormonal composition of the medium. The major classes of plant growth hormones are cytokinins, auxins, abscisic acid, gibberellins and ethylene. A higher concentration of cytokinins in the media usually favours shoot development and a higher concentration of auxins favours root initiation. The pH of the culture medium should be adjusted to the optimum level to favour the uptake of the nutrients from the medium in adequate levels. Low pH will affect the gelling of agar.

Maintenance of aseptic condition: Maintenance of aseptic condition is very important in tissue culture. Since the carbon sources are provided in the medium, the contaminants, such as bacteria and fungi will outgrow the explants and also produce toxic metabolites that would affect the growth of the explants. The different methods of sterilization are moist heat sterilization (Autoclaving), dry heat sterilization, filter sterilization and sterilization using radiation. Moist heat sterilization using an autoclave is usually adopted for sterilizing culture media. Glassware and tools are sterilised using dry heat or moist heat sterilisation. Filter sterilization is done for heat labile compounds, such as hormones, antibiotics, etc. The work area of the laminar air flow chamber is sterilized using UV radiation.

Routes of *in vitro* propagation: The major methods of micro-propagation are enhanced release of axillary buds, organogenesis and embryogenesis. Organogenesis and embryogenesis may be direct or indirect. In the indirect methods, an intermediate callus phase will be there, which is an unorganised mass of parenchymatous cells formed by the process called dedifferentiation. The process of regeneration of organs from the callus is called redifferentiation. Indirect methods often lead to variation in the progeny compared to the mother plants. Such tissue culture induced variation is called s**omaclonal variation.**

Tissue culture laboratory: A tissue culture laboratory should mainly be earmarked into a media preparation room, inoculation room and culture room. It should also have a hardening chamber for *ex vitro* establishment of the plantlets. The major equipments in the media preparation room includes autoclave, double glass distillation unit, electronic balance, pH meter, hot air oven, microwave oven, refrigerator, etc. Laminar air flow chamber is the major equipment in the inoculation room for providing sterile environment. The culture room should have illuminated shelves for keeping the cultures and air conditioners for the maintenance of temperature and humidity.

Problems of *in vitro* propagation: The main problems encountered during micro-propagation are contamination, vitrification, maturity barrier, polyphenol interference, loss during *ex vitro* establishment. Contamination can be overcome by the use of antibiotics, copper sulphate, etc. in the medium. Vitrification is a physiological disorder due to the absorption of excess water from the medium. It is a reversible phenomenon and can be overcome by modifying the C:N ratio of the medium, increasing the concentration of agar, etc. Decreased response of explants from mature tissues called as maturity barrier can be overcome by the use of hypomethylating agents. Darkening of the medium due to polyphenolic interference can be overcome by the use of compounds like activated charcoal, polyvinylpyrollidone, ascorbic acid, cold treatment, etc. Loss of plants during field establishment can be reduced by hardening under reduced light and increased humidity in mist chambers.

Plant tissue culture for crop improvement: Besides micro-propagation, plant tissue culture techniques can also be undertaken for crop improvement, such as production of novel plants by somatic hybridisation (fusion of protoplasts devoid of cell wall) and *in vitro* mutation. Production of homozygous lines through haploid culture can reduce to time taken for traditional breeding. Production of seedless fruits by endosperm culture, production of secondary metabolites, production of plants with novel traits by exploitation of somaclonal variation, cryopreservation of endangered species etc. are other applications.

Commercial micro-propagation : Micro-propagation is being adopted in a commercial scale in various crops at the state and national level and the tissue culture industry is fast growing. There is also an increasing export demand for tissue culture raised plantlets. The major crop in which tissue culture plantlets are in high demand in Kerala is banana. As the tissue culture banana plantlets are initially free of viruses and other diseases and have been micro-propagated from high yielding mother plants, a high yield is usually obtained in the field also. Uniformity in harvest is another advantage. Other tissue culture propagated plants include orchids, gerbera, pineapple, teak, date palm, etc. In large scale commercial propagation it is mandatory to have the plantlets certified virus free by the National Certification System for Tissue Culture Plantlets (NCS-TCP) in laboratories accredited for the same by the Department of Biotechnology, Government of India.

References

Introduction to Plant Tissue Culture.

M.K. Razdan, Science Publishers, 2003. 375 Pages.

❑❑❑

Chapter 31

Climate Change

Dr. R.V. Manju (Professor) **and Manasa R.** (Research Scholar)

Department of Plant Physiology, Kerala Agricultural University, College of Agriculture Vellayani, Thiruvananthapuram-695 522

Global climate change is a change in the long-term weather patterns that characterize the regions of the world.Tackling this phenomenon is of utmost importance given the pivotal role that climate plays in the formation of natural ecosystems and the human economies and civilizations.

Causes of Climate Change

Greenhouse gases

These are gas molecules that have the property of absorbing infra red radiation emitted from Earth's surface and re-radiating it back to Earth's surface, thus contributing to the phenomenon known as the greenhouse effect. Carbon dioxide (CO_2), methane, ozone, chlorofluorocarbons and water vapour are the most important greenhouse gases, and they have a profound effect on the energy budget of the Earth system despite making up only a fraction of all atmospheric gases. Concentrations of greenhouse gases have varied substantially during Earth's history, and these variations have driven substantial climate changes at a wide range of timescales.

Human activities

Human hands on carbon dioxide emission via fossil-fuel combustion, deforestation in the form of forest fires, loss of wetlands are chief causal factors for weather extremes. The loss of forest cover, which normally intercepts rainfall and allows it to be absorbed by the soil, causes precipitation to reach across the land eroding top soil and causes floods and droughts. Paradoxically, lack of trees also exacerbates drought in dry years by making the soil dry more quickly. Human activities also give way to other greenhouse gases, such as methane (from rice cultivation, livestock, landfills, and other sources) and chlorofluorocarbons (from industrial sources).

Volcanic activity

Volcanic activity can influence climate in a number of ways at different time scales. Individual volcanic eruptions can release large quantities of sulphur dioxide and other aerosols into the stratosphere, reducing atmospheric transparency and thus, the amount of solar radiation reaching Earth's surface and troposphere.

Climate Change affects Agriculture

The Agriculture sector is highly dependent upon weather and climate. Thus, it is vulnerable to climate change. Agriculture is not only sensitive to climate change but also one of the major drivers for climate change. The climate sensitivity of agriculture is uncertain, as there is regional variation in rainfall, temperature, crops and cropping systems, soils and management practices.

Agricultural productivity is affected in two ways:

1. Direct effects of increasing carbon dioxide concentrations.
2. Indirect effects of increasing concentration of CO_2 induced temperature buildup, precipitation, changes in soil moisutre and altered distribution and frequency of infestation by pests and diseases.

Impacts on plant systems

Crop yields are strongly affected by temperature and precipitation extremes along with CO_2. One additional factor that alters crop yield is CO_2, which stimulates growth by means of higher net photosynthetic rates for some crops (C_3 crops like cotton, wheat and rice) but has small growth effects on C_4 crops (sugarcane, corn, sorghum and millet) except under drought. Higher CO_2concentrations also reduce transpiration as plants reduce their stomatal apertures.

Our knowledge of plant responses to future CO_2 concentrations rests on the results of experiments that have experimentally increased CO_2 and then by comparing the performance of the experimental plants with those grown under current ambient CO_2 conditions. Such experiments have been performed in a wide variety of settings, including greenhouses and chambers of a variety of sizes and designs, like Open Top Chamber (OTC) and Free Air CO_2 Enrichment (FACE), Temperature Gradient Tunnel (TGT), Free Air Temperature Enrichment System (FATE) controlled environmental chambers.

Warming will accelerate many microbial processes in the soil-flood water system, with consequences for the C and N cycle. Crop residue decomposition patterns may change. Increased soil temperature may also lead to an increase in autotrophic CO_2 losses from the soil caused by root respiration, root exudates, and fine-root turnover.

Weeds, pests and diseases

Weeds evolve rapidly to overcome control measures, short-lived weeds and those that spread vegetatively evolve at the greatest rate owing to carbonaceous fertilisation. With changes in temperature and humidity levels, the populations of pests and pathogens may expand their geographic range, and expose animals and humans to diseases to which they have no natural immunity. A growing number of pests and diseases could lead to higher and even unsafe levels of pesticide residue and veterinary drugs in foods.

Crop Biodiversity

The distribution of wild crop relatives, an increasingly important genetic resource for the breeding of crops, will be severely affected leading to fragmentation of the distribution and even extinction.

Livestock

Climate change can alter livestock growth, reproduction rates, death rates, feed supplies and nutritional content disease and pest incidence. Heat waves have been found to affect animal performance, animal mortality, illness, feed intake, rates of gain, milk production, conception rates and appetite loss. Feed availability and feed quality are altered through effects on crop, pasture, and forage growth and nutritional quality, as discussed above. Increased rainfall intensity has been observed to increase range and pasture land soil degradation.

Incomes, income distribution and food security

Moderate warming and more carbon dioxide in the atmosphere may help plants to grow faster. However, more severe warming, floods, and drought may reduce yields, distressing welfare, food security and incomes. Moreover, it may affect poor people's livelihood through its effects on health, access to water and natural resources, and infrastructure.

Mitigation and Adaptation in Agriculture

- Assist farmers in coping with current climatic risks by providing value-added weather services to farmers. Farmers can adapt to climate changes to some degree by shifting planting dates, choosing varieties with different growth duration, or changing crop rotations.
- An early warning system should be put in place to monitor changes in pest and disease outbreaks. The overall pest control strategy should be based on integrated pest management because it takes care of multiple pests in a given climatic scenario.
- Participatory and formal plant breeding to develop climate-resilient crop varieties that can tolerate higher temperatures, drought and salinity.
- Selecting genotype in crops that have a higher per day yield potential to counter yield loss from heat-induced reduction in growing periods.
- Preventive measures for drought that include on-farm reservoirs in medium lands, growing of pulses and oilseeds instead of rice in uplands, ridges and furrow system in cotton crops, growing of inter-crops in place of pure crops in uplands, land grading and leveling, stabilization of field bunds by stone and grasses, graded line bunds, contour trenching for runoff collection, conservation furrows, mulching and more application of Farm yard manure (FYM).
- Efficient fertilizer use, such as optimum fertilizer dose, split application of nitrogenous and potassium fertilizers, deep placement, use of *neem, karanja* products and other such nitrification inhibitors, liming of acid soils, use of micro-nutrients such as zinc and boron, use of sulphur in oilseed crops, integrated nutrient management.

- Provide greater coverage of weather linked agriculture-insurance.
- Intensify the food production system by improving the technology and input delivery system.
- Develop a long-term land-use plan for ensuring food security and climatic resilience.
- National grid grain storages at the household/community level to the district level must be established to ensure local food security and stabilize prices.
- Provide incentives to farmers for resource conservation and efficiency by providing credit to the farmers for transition to adaptation technologies.
- Provide technical, institutional and financial support for establishment of community banks of food, forage and seed.
- Provide more funds to strengthen research for enhancing adaptation and mitigation capacity of agriculture.

From the above, it is clear that despite the fact that increase in atmospheric CO_2 are projected to stimulate growth and improve water use efficiency in some crop species, climate impacts particularly heat waves, droughts and flooding, have its foot print on crops, livestock, pests and their variability is to be expected to dampen yield potential. Consequently, climate change is certainly an agricultural issue portending negative effects in some regions, positive effects in others. Their adverse impact on livelihood of farmers is tremendous. Agricultural scientists need to guide farmers on projected impact of climate change and sensitise them on probable mitigation and adaptation options to minimize the risk in Agricultural sector.

References

Gepts, P., Izaurralde, C., Nelson, G., and Rosenzweig, C., 2016. How will Climate Change Affect Agriculture?. *Am. Assoc. Advmt. Sci.* 55p.

Ghosh, P., 2009. Is India a Solution to the Problem or a Problem to the Solution?. Lasting Solutions for Development Challenges. In: Narain, S., Ghosh, P., Saxena, N. C., and Soni, P. (eds), *Climate Change perspectives from india,* United Nations Development Programme (UNDP), 72p.

Mahato, A., 2014. Climate change and its impact on agriculture. *Int. J. Scientific Res. Publ.* 4(4): 1-6.

Taub, D., 2010. Effects of Rising Atmospheric Concentrations of Carbon Dioxide on Plants. *Nat. Educ. Knowl.* 3(10): 21.

□□□

Chapter 32

Rainshelter - A Low Cost Technology for Off-season Vegetable Cultivation

Dr. Sajeena, S. (Asst. Professor) and
Dr. Abdul Hakkim, V.M. (Professor)

CoH Padannakkad

Introduction

The changing agricultural scenario of Kerala encourages high investment and today, horticulture is moving from rural confines to commercial ventures. This change has proved to be intellectually satisfying and economically rewarding and hence, has started attracting educated youth. In addition to the efficient utilization of natural resources, horticultural crops are highly remunerative also. Ninety five percentage of vegetable production of Kerala is during summer. In spite of water availability, vegetable cultivation is negligible during rainy season due to high intensity rainfall distributed over two monsoons from June to November. Kerala has a unique climatic condition and the humid tropical condition restricts effective adoption of protected cultivation techniques followed in different parts of the Country. A greenhouse with sophisticated environmental control is not essential in Kerala and the climatic as well as economic constraints preclude its acceptability to small and marginal farmers of the State. The most effective alternative is the RAINSHELTER- a low cost technology for protected cultivation, which facilitate off-season production of vegetable crops and cut flowers. This technology is very acceptable to small scale and marginal farmers and can provide self-employment for the literate and unemployed youth.

Specification of Rainshelter

The design of a rainshelter affordable to small and marginal farmers of Kerala is shown in Fig. 1 and estimate of a rainshelter of

100 sq.m area is given in Table 1. A typical rainshelter suitable for Kerala condition is shown in Plate 1. The essential features of such a rainshelter can be summarised as follows:

Site

- An open area with ample sunlight
- Orientation should be in the North–South direction
- Irrigation and drainage facilities should be available

Structure

- Framework with GI pipes.
- Cladding with U. V. stabilized polythene sheet (200microns)
- Sides must be kept open
- Sharp edges should be avoided in the framework.

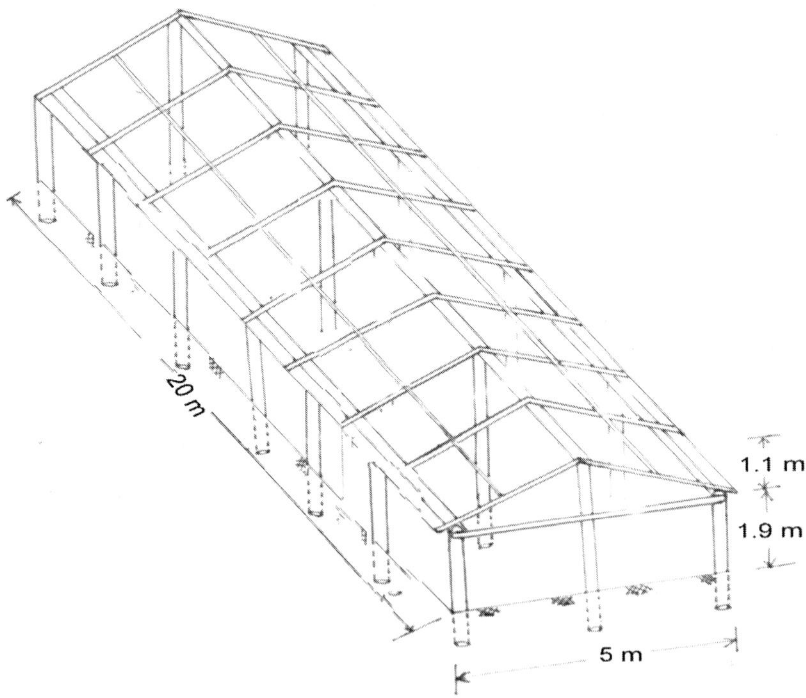

Fig. 1: Schematic Diagram of Gable Shaped Rainshelter (100 sqm)

Plate 1. Typical rainshelter suitable for Kerala condition

Table 1. Estimate of 100 sq.m with GI and MS pipe structure

Material	Quantity	Rate (Rs)	Amount (Rs)
1" GI pipe (B class)	160 m	90/m	14400.00
1.5"GI pipe (B class)	12 m	130/m	1560.00
½" GI pipe	25 m	60/m	1500.00
9 mm flat (MS)	80 kg	60/kg	4800.00
20mm bolt & nut-60cm long	90 kg	70/kg	6300.00
U.V. sheet	160 sq.m	70/sq.m	11200.00
Shade net	140 sq.m	24/sq.m	3360.00
Laterite stone	180 Nos	30/no.	5400.00
Welding charges			3000.00
Labour charges			10000.00
Concrete work	1.5cu.m	4000/cu.m	6000.00
Drip irrigation system	100 sq. m	100/sq. m	10000.00
Miscellaneous			2480.00
Total			**80,000.00**

The introduction of "rainshelter" for off-season vegetable cultivation can help to increase the income of farmers and lead to achieve self-sufficiency in vegetable production. Vegetables like tomato, salad cucumber, bhindi, capsicum, etc. can be cultivated very successfully during off-seasons.

Conclusion

Rainshelter is a low cost technology for protected cultivation which facilitate off-season production of high value crops like tomato, capsicum, cut flowers, etc. This technology is acceptable to small scale and marginal farmers and this type of intensive cultivation creates self-employment for the literate and unemployed youth of Kerala.

References

Bongiovanni R and Lowenberg-Deboer J., 2004. Precision agriculture and sustainability. *Precision Agriculture,* **5**: 359-387.

Van Schilfgaarde J., 1999. Is Precision agriculture sustainable? *American Journal of Alternative Agriculture,* **14(1)**: 43-46.

❑❑❑

V. Agriculture and Allied Activities

Chapter 33

Value Addition of Fruits and Vegetables-Prospects for Entrepreneurs

Dr. Mini C. (Professor & Head)

Department of Post-Harvest Technology, Kerala Agricultural University, College of Agriculture Vellayani, Thiruvananthapuram-695 522

India is the second largest producer of fruits and vegetables in the world. Unlike other agricultural crops, the horticultural crops, especially fruits and vegetables are highly perishable in nature with limited shelf life. It is a paradox that in spite of the fact that India is the leading producer of almost all horticultural crops, the post-harvest losses in our country are among the highest (20-40%) in the world due to gaps in the cold chain, such as poor infrastructure, insufficient cold storage capacity, unavailability of cold storages in close proximity to farms, poor transportation infrastructure, etc. The problem is severe during "glut" period because of the inadequate availability of post-harvest management technologies. The need of the situation is to make fruits and vegetables available for consumption throughout the year in processed or preserved form. This necessitates development of an effective value chain starting from production to consumption that in turn will give higher returns to the growers and entrepreneurs. Since fruits and vegetables are highly perishables, different types of processing and value addition technologies are required to preserve them.

The scenario of fruit and vegetable processing industry in India is fast changing. Since nineties there has been a constant flow of funds into this sector and at present, the new industrial policy has placed fruit and vegetable processing in the list of high priority area. The food processing industry is now identified as a "sunrise industry" owing to its enormous input potential and its significance in the development of the country. Though the number of fruit and vegetable processing units, their installed capacity and production have grown up steadily in the last few years, fruit and vegetable processing industry in India is still in a nascent and primitive stage. Number of establishments in the organized sector is less, when compared with several developed and developing nations. The technologies adopted for processing and value addition in medium and small industries are primitive and outdated. Processing of fruits and vegetables into various products using proper cost effective technology is a viable tool for improving economic status of farmers and our country.

Processing Technologies-Traditional and Modern

Demand for a particular processed form, quality and regular supply of the raw material are some of the major factors, which determine the economic feasibility of processing. The technology for preservation varies with type of products and targetted market. There are various methods of fruit and vegetable preservation including thermal processing, fermentation, pickling, dehydration, freezing, etc. Value addition through processing can be brought out by converting them to various products like fruit beverages and sugar-based products, most of which do not require large investment in machinery and can be set up in a small scale sector. Fruit juice, RTS beverage, squash, syrup, cordial, nectar, etc. constitute fruit beverages. Fruit jam, jelly, marmalade, preserve, candy, fruit bar, toffees, etc. are the common sugar based products. Chutneys and sauces are the other specific products that can be tried by processors. Modifications in the usual traditional technologies can bring novel ideas in product development improving consumer acceptance and enhanced returns. Use of fruit juice in carbonated drinks enhances nutritive value, juice concentration reduces bulkiness, storage and transportation cost, and fruit juice powder production avoids inconvenience of transporting large volume of liquid.

With the increasing health consciousness of consumers, there has been an increasing concern over the use of thermal processing methods like blanching, pasteurization , sterilization, etc. for enhancing the shelf life of foods, as these methods fail to preserve the fresh like nutritional, therapeutic and sensory qualities of the food. There has been a shift from the traditional heat processing to those technologies, which employ no or minimum heat application, thereby retaining maximum freshness and quality. Over the past few years, a number of non-thermal processes have been developed for extension of shelf life. These techniques are now being employed to meet consumer trends mainly by their use in combinations. Irradiation, high hydrostatic pressure processing, extrusion, ohmic heating, microwave processing, pulsed light treatment, use of ultrasound, etc. are some of the important novel advanced technologies that can be employed in fruit and vegetable processing. Hurdle technology is a new concept developed to ensure safe and stable food by employing an intelligent simultaneous or sequential combination of different preservation factors or techniques (hurdles) to achieve multi- target, mild but reliable preservation effects.

Quality foods with fresh and fresh like attributes are highly demanded now-a-days to meet the changing trends of modern consumers and interest in minimal processing is increasing accordingly. Consumers have become more health conscious in their food choices but have less time to prepare healthy meals. Minimal processing technologies provide sufficient shelf life to foods to allow their distribution and also meeting the demands of the consumers for convenience and fresh-like quality. Minimally processed foods are fresh raw produce, processed in order to supply a ready-to-eat or ready-to-use products.

Packaging: An Integral Part of Processing

Packaging has an enormous impact on food preservation. Packaging is the key to successful marketing of modern food products. Apart from extending shelf life and aesthetic appeal, packaging provides convenience of size, shape and ease of opening. Packaging materials with appropriate barrier properties are chosen to design suitable package for foods depending on their physical nature and to preserve the quality and nutritive value of foods. Packaging also prevents spoilage of food by microbial or insect attack. Proper selection of the package depends on product characteristics, their nature of deteriorative changes, conditions like physical hazards in handling, transportation, storage and climatic conditions (RH and temperature) that brings out changes. They should not be over or under packed, as over-packaging increases packaging cost and under-packaging fails to offer desired protection. Hence, a functional and economic package is desirable for the processed fruit and vegetable products.

Traditional packaging is only suitable provided the climate does not cause an increase in the moisture content of the food, which will result in mould growth. Boxes are used to prevent crushing of dried foods, and in humid climates, moisture-proof flexible films can be used. Some semi-moist foods, such as osmotically dehydrated fruits have special needs to prevent the re-absorption of water. A wide range of flexible packaging materials are also available, but the use of many of these is limited due to high cost. Low-density polyethylene is a moderately good moisture barrier, cheaper than other films and easily sealable. Flexible materials may be used as the sole component of a laminated package, but for most foods, a sturdy outer container is also needed to prevent crushing or to exclude light.

Products that are packaged inappropriately or with uninformative labelling, regardless of the quality have less selling and are against the packaging and labelling regulations of FSSAI. Jam, chutney, and ready-to-serve beverages are commonly packaged in glass bottles, while higher-volume fruit juices are packed in plastic cans as well. The use of attractive, internationally competitive packaging for local products would represent a successful marketing strategy. The size of the product is also important. Most consumers prefer single-serving products, or sizes that can serve 4-6 persons at the household level.

Product Quality; An Important Tissue in Global Food Supply Chain

Quality may be equated to meeting standards required by a selective customer. To produce high quality products, processor needs to be aware of quality attributes, which consumer discerns as most important and relevant in determining acceptability. Most consumers usually judge product acceptability based on their appearance, flavour, texture and perceived nutritional benefits. These attributes are functions of physico- chemical composition of fruits or vegetables, which are further influenced by several factors viz., quality and composition of raw materials, processing techniques, effect of environmental factors encountered during storage and distribution, customer handling and protection provided by packaging. Food safety is another important factor influencing quality of processed fruits and vegetables.

Use of sound fruits and vegetables of correct maturity, adoption of proper and hygienic processing practices, maintenance of personal cleanliness, proper education and experience for the persons monitoring sanitation programmes, design and construction of processing plant with enough space for sanitary arrangement, use of simple, efficient and self-cleanable or easily cleanable equipments, facility to conduct physical, chemical and microbiological quality analysis, etc. are to be assured for producing and maintaining good quality processed products.

Conclusion

Only 2-3% of total production of fruits and vegetables produce are processed in India, which is extremely low as compared to the International averages. More and more customer demand and increasing global competition are pressurizing to improve the pace and quality of innovations in processing sector. Increasing retailing and higher spending of consumer are expected to increase the penetration of processed food in the Indian market. The favourable agro climatic condition gives an immense export opportunity to the entrepreneurs of food processing. Entrepreneurship need more focus and emphasize for tapping all cashing opportunities. It plays a vital role in diversification and commercialization, enhances shelf life, ensures value addition, generates employment, enhances farmer's income and creates market for export. However, there are lots of barriers in the entrepreneurship of fruits and vegetable, processing pertaining to technology, finance, marketing and export. There is an urgent need to provide facilities for entrepreneurs for running an economically viable unit.

References

Sudheer, K.P. and Indira, V., 2018. *Entrepreneurship and skill development in horticultural processing.* New India Publishing Agency, New Delhi. 432 p.

Mini , C., Geethalekshmi, P.R. and Manjunath Shetty, 2018. Underutilized fruits and vegetables- Scope for value addition. pp. 161-188. In. Sudheer, K.P. and Indira, V. *Entrepreneurship and skill development in horticultural processing,* New India Publishing Agency, New Delhi. 432p.

□□□

Chapter 34

Jackfruit products suiting urban lifestyles

Dr. Suma Divakar (Professor)

Department of Community Science, Kerala Agricultural University, College of Agriculture Vellayani, Thiruvananthapuram-695 522

Jackfruit is a prominent food crop of Kerala Jackfruit tree, though not grown in plantations, is an inevitable component in most of the homesteads. Today, jackfruit is considered as the most organically grown and unpolluted fruit crop. Jackfruit is well known for its ethnic taste and is used in various cuisines in different countries. Considering the scope of the edible components of this fruit, it is also an answer to the challenge of food security of the state. Thus, it has been declared as the 'State fruit of Kerala'.

Jackfruit (*A. heterophyllus*) is a rich source of vitamins, fibre and minerals. It can be incorporated into food products to increase their nutritional content. It possesses a wide variety of nutritional components, such as Carotene, Vitamin B_1, Vitamin B_2 and Potassium, Fructose, Glucose and Sucrose are found to be the major sugars contained in them. Capric, myristic, lauric, palmitic, oleic, stearic and linoleic acids are the major fatty acids found, in varying proportions in the different parts of jackfruit (Chandrika *et al.*, 2004).

Invariably, the production season coincides with monsoons, making the fruits unacceptable in taste. Besides, the difficulties in harvesting, extracting the edible ripe fruits and attraction of flies are some of the main reasons constraining its wide scale use before and after rain (Prusky, 2011).

Besides, jackfruit is perishable and cannot be stored for a long time, because of its inherent compositional and textural characteristics. Every year, a considerable amount of jackfruit, specially obtained in glut season (June- July) goes waste due to lack of proper post-harvest knowledge during the stage of harvesting, transporting and storing, which affects both, the quality and quantity (Azad, 2000). Proper post- harvest technology for prolonging shelf life has, therefore, become necessary. Besides, alternate ways of using jackfruits in season can reduce post-harvest losses. Here in, processing technologies have a major role. It adds diversified and attractive food items to daily meals, as well as contributes to the generation of income and employment.

Changing lifestyles have made it all the more necessary to open new avenues for better utilization of jackfruit, as the traditional uses have become stablilized. Therefore, there lies great opportunities for the non-traditional uses of jackfruit, especially in the form of convenient foods. A few convenient foods standardized at the Department of Community Science, at the College of Agriculture, Vellayani are discussed herewith.

Jackfruit Based Curry Mixes

The dimensions of jackfruit bulbs and seeds were identified. It was then pre-treated to prevent browning. The 3 typical dishes of Kerala, namely 'avial' 'kootu' and 'ularth' were standardized from 5 formulations. These formulations were rated by a sensory panel comprising of 10 members.

'Avial' mix identified by the panel comprised of jackfruit bulbs and seeds, green chilly, garlic, cumin, turmeric powder and curry leaves. Similarly, 'kootu' mix comprised of jackfruit bulbs and seeds, red chilly, turmeric powder, cumin and curry leaves.'Ularth' mix comprised of jackfruit bulbs and seeds, crushed red chilly, onion, garlic, turmeric powder and curry leaves. The products were dehydrated in a drier at 65^{0}C till breaking stage (moisture content below 10%).

In order to give valuable hints to consumers for getting the best product, parameters like cooking procedures, cooking time and addition of adjuncts were also standardized. The shelf life of the products were found to be satisfactory for more than 3 months.

"Ulart h mix"

"Avial "mix

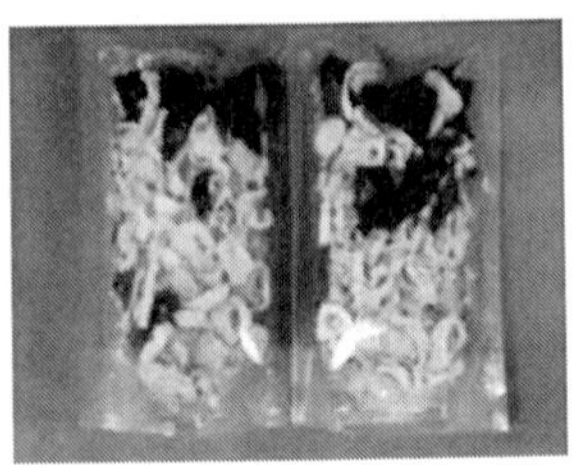

"Koottu "mix

Jackfruit based noodles

Flour of jackfruit bulb and seeds were prepared after pre-treatments. Bulbs and seeds were standardized for optimum width, blanching, immersion in different media, drying and milling. Composite flour was prepared by mixing refined flour, bulb flour and seed flour in different combinations (40:30:30, 50:25:25, 50:30:20, 50:40:10, 50:10:40, 50:20:30).

Water was added in phases to the composite flour and kept for refrigeration, so as to maintain the moisture level at 33 per cent. The moist flour was placed in Brabender Single screw food extracter and the temperature was set in different zones (55^{0}C, 45^{0}C and 60^{0}C). The dimensions set for extraction were 2mm X 4 strands. The extracted sheets were cut into strips. The noodles strips were dried in the Cabinet drier at 60^{0}C and sealed in HDPE and laminated covers.

Noodle formulations

Noodles were evaluated for their sensory parameters. Sensory scores indicated that combination T_5 (50:10:40) and T_6 (50:20:30) were highly acceptable.

Jackfruit based baked products

Two acceptable baked products were standardized using raw jackfruit bulb flour Rusk and Pizza base. The composite flour comprising of refined flour and jackfruit bulb flour in the ratio 80:20 gave products with excellent sensory qualities.

Straight dough method was adopted for baking. The fermentation time of the products were between 1-1/2 hours.The baking time for pizza base was 10 mts at 250^{0}C. Rusk was baked in 2 spells - 25 mts for baking into bread and 20mts for baking into rusk at 250^{0}C. When the shelf life of pizza base was 3 days, for rusk it was 3 months.

Jackfruit rusk

Jackfruit pizza base

Conclusion

The products standardized were all found to be more nutritious than commercial brands. Moreover, they were all chemical free. Thus, the scope for utilizing jackfruit proved to be successful. However, the products need scaling up before commercilisation.

References

Azad, A.K., 2000. Genetic diversity of jackfruit in Bangladesh and development of propogation methods. Ph.D. thesis, University of Southampton, UK.

Chandrika, V.G., Jangz, E.R. and Warnesuny, N.D., 2004. Analysis of carotenoids of ripe jackfruits in kerala and study the bio conservation. *J. Sci. Food. Agric.* 85(2): 186-190.

Madhavan, N., 1994. More fruits less processing. The Hindu daily. October 22nd. p. 6.

Prusky, D., 2011. Reduction of the incidence of post-harvest quality losses and future prospects.Food security 3(6): 463-474.

❑❑❑

Chapter 35

Introduction to Apiculture

Dr. Amritha V.S. (Asst. Professor), **Akhila Pahee** and **Chinchu P. Babu** (Research Scholars)

Department of Agricultural Entomology, Kerala Agricultural University, College of Agriculture Vellayani, Thiruvananthapuram-695 522

"If the bees disappeared off the surface of the globe, then man would have only four years of life left…no more bees, no more pollination, no more plants, no more animals, no more man…"

Albert Einstein

Kerala state is having a great potential for bee-keeping due to its diversified flora and honey bee species. The major nectar and pollen source to bees in Kerala are the rubber plant (*Hevea brasiliensis*is) and coconut palm (*Cocos nucifera*), respectively. Crops like teak, cardamom, cashew, coffee, sesamum and vegetables are also excellent sources of nectar and pollen. In addition, forest area of 10,81,509 ha with diversified flora are also available for bee-keeping. Various agricultural and horticultural crops are, thus, benefited by bee pollination.

Honey bees (Order: Hymenoptera, Family: Apidae) are social insects living in colonies with remarkable degree of social instincts and division of labour among the different members of the colony. The major honey bee species are as follows:

1. The Rock Bee : *Apis dorsata* Fabricius

Apis Dorsata F.

It is the largest of the honeybees. It builds open single comb of huge size about a meter in diameter. The comb is fully exposed and hung from inaccessible branches of trees, along the sides of steep rocks in the forest and even from walls, rafters and other parts of buildings. It produces plenty of honey and the annual yield from a colony is about 37 kg. It is impossible to domesticate because of its irritable and ferocious nature.

2. The Indian Bee : *Apis cerana indica* Fabricius

It is the common Indian honey bee found both, in the forests as well as in plains throughout the country. It is smaller than rockbee but larger than the little bee. It builds many parallel combs in the cavities of trees, caves and such other hidden sites. It is mild and capable of being domesticated and is commonly reared in South India. The annual yield of honey is 2-5 kg/colony. A queen can lay 350-1000 eggs per day. The hill bee race is larger in size and darker in colour than the plain race and has the habit of moving away from the broods even at slight disturbances.

3. The Little Bee : *Apis florae* Fabricius

It is known as the little bee since it is the smallest of the four species of *Apis*, found only in plains. It builds single, but small combs on bushy plants and corners of roofs. It yields very little honey, about 0.5-1 kg per year from a colony, it is not possible to domesticate. A queen lays 323-365 eggs per day.

4. The Italian Bee : *Apis mellifera* Linnaeus

In Europe and America, *A. mellifera* is extensively reared. It is very closely similar to *A. cerana indica* in structure and habit but has prolific queen and has good honey gathering qualities, yielding about 45-180 kg per year. It has now well-established in the Northern parts of the country.

5. The Dammer Bee : *Tetragonula iridipennis* Smith

The dammer bee is tiny with a vestigial sting and is different from the four species in appearance and habits. It inhabits crevices in walls and hollow trunks of trees. The comb is made up of a dark material called "cerumen", which is a mixture of wax and earth or resin. The brood cells are of the shape of tiny ellipsoid pellets, all joined together with bits of wax and resin. It is a very poor honey gatherer and yields only 60-180 ml per colony per year.

Castes of Honey Bees

The Bee Colony

In a bee colony of an average size, there may be about 20,000 to 30,000 bees consisting normally of a queen, a few hundred drones and the remaining individuals are the workers making over 90 per cent of the population.

Queen : The queen is functional mature female of the colony and only one is found in a normal colony. The only function she partakes is that connected with reproduction and maintaining the colony strength. Queen is able to lay both, fertilized and unfertilized eggs. Queen is raised in a special cell constructed at the lower border of the comb, which is much larger than others.The spermatozoa received during the mating are sufficient to last for her entire life period of 2 or 3 years. She lacks motherly instincts of rearing the larvae, which is done by nurse bees. When a queen becomes old and unable to lay fertilized eggs, another queen is reared and the old one is discarded.

Worker : The worker is imperfectly developed female and is the smallest of the three. The workers are selfless members of the colony devoting all their attention and time to the service of the colony as a whole. The workers are raised in the cells (with flat caps) situated in the lower part of comb. The adult worker bees live for 45-80 days.

Drone : Drone is the functional male raised from unfertilized eggs and is larger, darker than the worker. It has no sting, wax glands or pollen baskets. In a colony, only few hundred drones or no drones may be found. Its only duty is to take part in nuptial flight of the queen and to mate with her. The drone cells are covered by a convex cap with a central hole. Longevity of drone is about 60 days.

Bee-keeping Equipments

Honeybees are reared in artificial hives for the produces they give i.e., honey, bees wax, propolis, royal jelly, etc. The practice of rearing bees is called Apiculture. The place where the hives are maintained is called an Apiary.

- Bee veil- It is worn over the face during handling of bees in order to protect our face from stings. It should be made of light material, such as nylon netting.
- Smoker - It is used to give smoke to the bees for easy handling by masking the alarm pheromones, thus, reducing the chances of being stinged by them. It consists of a tin can provided with a spout for directing smoke from the smoldering material inside it with the help of bellows.
- Queen cage- It is used to cage the queen and for carrying queen from one place to another during colony division or collection of swarmed colonies.
- Hive tool- It is a piece of flattened iron. It is used for prying apart the frames in the hive and for scraping bee glue and superfluous pieces of comb from various parts of hive.
- Bee Brush- A bee brush or a whisk broom is often employed to brush off bees from honey combs, before it is taken for extraction.
- Queen Excluder Sheet- A perforated sheet of zinc or round wires assembled together 37.50 mm apart for the species *Apis cerana indica.* It is inserted in super chamber to prevent the entry of the queen and thus, preventing it from laying eggs in the super chamber.
- Honey Extractor- The equipment consists of a cylindrical drum containing a rack or box inside to hold the super frames. The honey in the cells is forced out in droplets by the action of centrifugal force and can be collected in vessels through an exit in the drum. The combs after extraction can be used again.
- Bee Knife- When the cells of honey comb have been filled with honey, they are sealed by the worker bees. These wax capping have to be removed with the help of bee knife before the honey extraction.
- Comb Foundation Sheet- Made of pure wax, it is artificially provided for the colonies during honey flow seasons by cutting them to a proper size and attaching them to super frames by means of thread or fiber, the bees construct super structure of comb cell over the sheet.

- Feeder- It is used to provide sugar syrup to feed the honeybees (artificial feed).
- Swarm Catching Equipment- The swarm can be collected in a swarm catching basket by gently directing the bees into it.
- Pollen Trap- It is fitted over the hive entrance and as the returning foragers passes through these openings, pollen pellets collected in the hind legs gets trapped in it.
- Comb Foundation Mill- Comb foundation mill is used to print the comb into the flat sheet.
- Queen gate- It is a piece of queen excluder sheet fitted on slotted entrance gate. It confines queen inside the hive and is used to prevent swarming and absconding.

Bee hives

For rearing of bees, first of all we need a bee hive. Bee hive is a man-made structure in which the bee colony lives. There are different types of hives used for hiving bees like :

- Newton's hive A&B : Indian bees
- Langstroth hive : Italian bees

Honey bee colonies should be opened, checked, or manipulated one to four times each month from February to October. Timely management practices are required for productive bee keeping.

Seasons in Bee-keeping

There are three different seasons for bee-keeping in Kerala. They are:

1. Dearth period (June-August)

In this period, nectar and pollen are not available to the bees. Egg laying activities decreases or stops, as there is no food to feed the brood. During dearth season:

- Provide sufficient shade to the hive by bee-keeping it under trees or artificial structure.
- Provide artificial feed (sugar syrup), pollen supplement and provision for clean water. (Pollen supplement is given only in

north Indian conditions). In our conditions during dearth, the hives are kept in coconut orchards to get sufficient pollen and nectar.

Preparation of Artificial Feed

Boil the water in a clean vessel. When water gets boiled, add sugar slowly by continuous stirring and allow it to cool. Add 2 g of turmeric powder and feed the colonies in a feeder. Take care not to prepare feed in open condition in the apiary and avoid dripping to prevent robbing by ants and bees. Feed the colonies starting from one end and not in random.Shallow feeder/coconut shells are used to feed. The feed rate for Indian bee is 250g/colony/week and for Italian bee is 750g/colony/week. If population is weak, reduce feed quantity to 200g and 700g, respectively. During brood rearing season, the interval can be reduced in between the feed (between 2-3 days).

2. Brood Rearing Season (September-December)

During this period, all the stores are used for comb building, egg laying and brood rearing. At this time, there should be more space in the frames for the queen to lay eggs to maximum capacity and brood rearing may not drop. During this period:

- Replace old combs which are dark/wax moth infested with new ones.
- Feed colonies more frequently so that queen bee lays more number of eggs and more workers are there for honey flow season.
- Division of colonies is done to prevent swarming (natural division of colonies, when population increases) of colonies. Division is usually done during evening hours.
- Empty frames should be placed between combs, comb foundation sheets are also provided to speed up the comb building process.

Division of Honey Bee Colony

During the growth period, the queen bee lays more number of eggs and hence, there will be more workers and also drones. Then, there comes new queen cells in the colony. Before the hatching of new queen cell, the old queen takes almost half of the bees and forms a new colony. Thus, the colony divides naturally by swarming. We have to divide the colonies in order to prevent swarming.

3. Honey Flow Season (January - April)

This season coincides with spring. During this season:

- Provide super chamber during December-January and provide more space for honey storage by adding four frames in super.
- Confine queen to brood chamber using queen excluder sheet between brood chamber and super chamber and prevent swarming.
- Ensure sufficient population, combine weak colonies with strong ones.

Pests and Diseases

Pests:1. Varroa mites (*Varroa destructor, Varroa jacobsoni*): Brood infested with a few adult mites emerge without visible damage.The symptoms include loss of haemolymph, deformed wings and legs and spotty brood pattern with punctured pattern.

2. Greater wax moth (*Galleria mellonella*): Larvae destroys comb by burrowing through them, constructing tunnels of silk and devour the combs for protein in pollen, cocoons and propolis. Attack the midrib section of the comb & often reduces the central portion to a mass of silken web & frass.
3. Wasp (*Vespa cincta*): The wasp waits near hive entrance and catches bees coming out and feed them.

Other pests include lesser wax moth (*Achroia grisella*), small hive beetle (*Aethina tumida*), ants, etc.

Diseases

1. American foul brood (*Paenibacillus larvae*): The caps of dead brood cells are usually darker than the caps of healthy cells, sunken, and often punctured. Presence of scattered, irregular pattern of sealed and unsealed brood cells. Sunken and punctured cappings are the common symptoms. Stretch test is used to detect the disease.
2. European foul brood (*Mellissococcus pluton*): The diseased larvae turn yellow and then brown and the tracheal system becomes visible. Larvae die in a coiled stage causing a foul smell.

Other diseases include Nosema disease (*Nosema apis*), Chalk brood (*Ascosphaera apis*), stone brood (*Aspergillus fumigatus, A. flavus*), etc.

3. Thai sac brood virus disease: Larval skin is quite tough and has the appearance of a small, watery sac.Dead larvae remaining within their cells eventually dry out to flat scales that adhere loosely to the cell floor.

Augmentation and conservation of honeybees will definitely be benefitted to the farming community, which will enhance crop productivity through pollination by bees. In cardamom plantations, honey bees are the sole agent of pollination. Coconut, cashew, coffee, vegetables are the other crops benefitted by bee assisted pollination which in turn supplement for yield enhancement to a tune of 20-40 per cent. Food security, food diversity and productivity enhancement can be ensured through bee-keeping in all the districts. Bee-keeping may be considered as an essential component of crop husbandry practices and thus, an integral part of agriculture.

References

Abrol, D.P., 2000. Beekeping with *Apis cerana* in Jammu & Kashmir: Present status and future perspectives. *Bee Wld*: 149-152.

Atwal, A.S., 2006. The World of the Honey bee, Kalyani publishers. 115pp

Devanesan, S., Prathapan, K., Premila, K.S., Shailaja, K.K. and Amritha, V.S., 2011. Bee-keepers orientation training manual p. 75

Mishra, R.C., 1995. Honey bees and their management in India, Indian Council of Agricultural Research, New Delhi, pp. 112-115.

❑❑❑

Chapter 36

Sericulture

Hari Sankar S.S. and Gayathri P. Bose (Research Scholars)

Department of Agricultural Entomology, Kerala Agricultural University, College of Agriculture Vellayani, Thiruvananthapuram-695 522

Sericulture is the science of rearing of silkworms for harvesting silk from them. It is derived from the French word "Seris" meaning silk and cultura means raising. Silk was extracted from silkworm as early as 2000 BC in China. Most commercially exploited species for silk production is Mulbery silkworms *Bombyx mori*. Other than this in India, few other species are commercially reared like Eri silkworm (SW) (*Samia cynthia ricini*), Tassar SW (*Antheraea mylitta*), Muga SW (*Antheraea assamensis*).

S. No.	Common name	Scientific name	Extend of domestication	Silk character	Host
1.	Mulberry SW	*Bombyx mori*	Domesticated	Single continuous reelable thread (500- 1500m)	Monophagous *Morus* spp
2.	Eri SW	*Samia cynthia ricini*	Domesticated	Discontinuous, non-reelable silk	Polyphagus, *Ricinus communis, Manihot esculentus,* Champa
3.	Tassar SW	*Antheraea mylitta*	Wild	Dense, Firm, continuous thread with supporting loop/peduncle on cocoon	Polyphagus, *Shorea robusta, Zyzipus zuzuba, Tectona grandis, Terminalia* spp
4.	Muga SW	*Antheraea assamensis*	Semi domesticated	Reelable silk. Rudimentary peduncle in cocoon. Glossy cocoons	Polyphagus *Machilus bombicina, Litsea polyanthe*

Mulberry SW,
Bombyx mori

Eri SW,
Samia cynthia ricini

Tassar SW,
Antheraea mylitta

Muga SW,
Antheraea assamensis

Mulberry Silkworm Rearing

Eggs are collected from certified grainages and black boxed for uniform emergence. The emerged first instar larvae are called chawki. They are fed with finely chopped tender leaves of mulberry. There are two types of rearing system based on feeding type. They are shoot rearing; entire shoots are fed and larvae are reared on beds, and leaf rearing; where larvae are fed with cut leaves and reared on trays made of wood. The trays are lined with wet foam rubber strips and butter paper to ensure high humidity for the early instars. Bed cleaning should be done daily until third instar, on alternate days for final instars.

As the larvae progress in size, they are appropriately shifted to newer trays to avoid crowding. The feeding pattern also changes with instar. Later instars should be fed with mature and coarser leaves. Final instar larvae ready to moult are called ripe worms, and such worms are transferred to structures that facilitate spinning cocoons without crowding called as mountages. They are either made of bamboo or plastic. Locally they are called as chandrikes, netrikes, etc. ripe worms are identified by their pale yellow colour and characteristic head spinning behaviour, indicating readiness to spin cocoons. Two to three days after all worms are transferred to mountages, the cocoons are ready for harvest. Harvested cocoons are sterilized to prevent adult emergence and facilitate storage, through a process called stifling. They are either boiled in water, dried in hot air oven, or sun-dried.

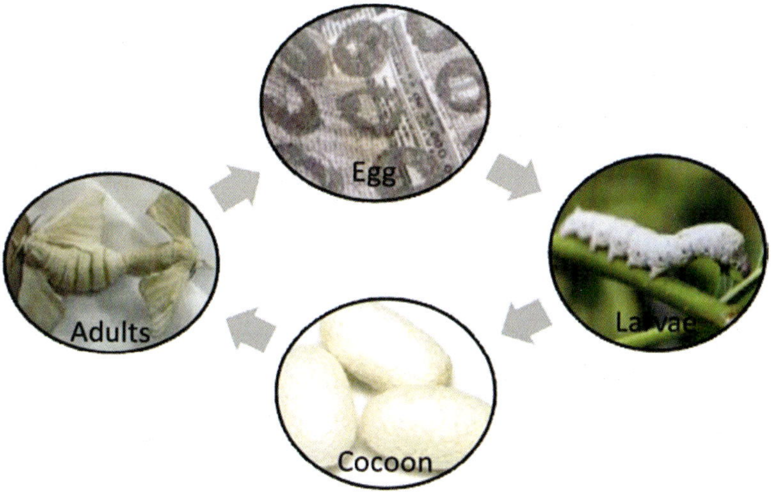

Stifled cocoons are ready for reeling. To remove the outer sericin, they are boiled and loose end of the thread is identified. It is then attached to a reeling unit and reeled to fibres.

Common diseases and pests of silkworm

S. No.	Disease/ pest	Causal organism/ SN	Symptoms/damage	Management
1.	Pebrine	*Nosema bombycis*	Pepper like black spots on larvae, unequal development of worms, rubbery skin. Transovarial transmission	Mother moth examination, proper disinfection, bed hygiene
2.	Flacherie	Bacteria	Flacis larvae, shrunken body, diarrhoea, foul smell	Bed hygiene, proper nutrition, use of bed disinfectants
3.	Muscardine	*Beauveria bassiana* (fungus)	Mummified chalk like larvae	Do not provide wet leaves, maintain dry beds
4.	Grasserie	Nucleopolyhedro virus (NPV)	Larvae becomes brittle, milky fluid oozes out	Bed hygiene, proper nutrition, use of bed disinfectants
5.	Uzi fly	*Exorista sorbillans*	Parasitoid of SW, adult emerge by breaking cocoon, rendering it unreelable.	Clean cracks and crevices, fumigation after each season, provide mesh over window pane

Common disinfectans used in sericulture

Room disinfectants

- Formaldehyde - 4%
- Bleaching powder - 5%

Bed disinfectants

- Vijetha, Ankush, Astra, Resham Keet Aushadh (Formaldehyde + Benzoic acid + lime powder).

Skill development scheme for sericulture

Central Sericulture Research and Training Institute (CSRTI) Mysore offers various training programmes for technical staff and budding entrepreneurs. Training programmes of varying duration, to instill a sense of confidence and equip themselves with the necessary skills required to successfully take up sericulture as a profession is offered by CSRTI. (www.csrtimys.res.in/training/structured).

Sericulture opportunities in Kerala

Kerala, being a land of bountiful resources, could nurture and flourish sericulture as a successful entrepreneurship. A family of four can easily manage a sericulture enterprise in their backyard. It is advisable to grow your own mulberry. 5-10 cents of backyard land can easily accommodate a rearing shed and mulberry garden. High yielding mulberry varieties can be procured from Mysore and planted in rows. They can be pruned to appropriate height to facilitate easy harvest. Normal yields range from 6 to 8 tons of leaves per year per hectare. The silk waste like dead larvae, unopened cocoons, reeling wastes, etc. are excellent manure, fish and poultry feed. They are also used for industrial oil extraction and the oil meal is also an excellent manure. Infrastructural development demands capital, which may be sourced through bank loans. Legislative intervention is necessary in Kerala to subsidize sericulture related enterprises. Government of Kerala recognizes sericulture as a non-traditional activity under the Rural Development Department. The production and productivity is on the rise in our state, recording 31 MT during 1995-96 to 101 MT in 2006-2007, the highest in the sericulture history of the state. However, State cocoon production in 2013-14 was only 40 tons.

The major components of Sericulture schemes as per Rural development department are :

- Support for mulberry plantation and silk worm rearing.
- Support for extension and publicity.
- Beneficiary empowerment programme.
- Incentive for Bivoltine silk yarn production.
- Production incentive for quality linked cocoon procurement system.
- Support for mechanisation of existing mulberry garden.
- Support for maintenance of existing mulberry garden.
- Support for post-cocoon sector (Reeling, twisting, weaving, printing, handicraft, etc.)

Subsidy

- Plantation subsidy - Rs. 6750/ Acre
- Equipment subsidy - Rs. 37500/Acre
- Rearing shed subsidy - Rs. 1,00,000/Acre
- Irrigation subsidy - Rs. 30,000/Acre
- Production incentive - Rs. 30-50/1 Kg cocoon
- (GC+% 80 to 90 and above)
- Mechanisation of Mulberry garden - Rs. 30,000/Acre
- Maintenance of existing Mulberry garden - Rs. 4150/Acre

[Data as per rdd.kerala.gov.in]

Women self-help groups like kudumbashree units could successfully run sericulture enterprises. Through appropriate interventions, the Government could devise schemes to popularize sericulture and related industries, thereby attracting youth and jobless homemakers to take up sericulture in small scale. There arises a need for cocoon markets as well in Kerala, which would facilitate the smooth run of Industries. Famers could also tie up with weaver towns in the neighbouring states like Kancheepuram in Tamil Nadu to outsource their quality silk. Kerala being a tourist hub could attract flocks of tourists to sericulture rearing houses, silk reeling and saree making industries under one roof, and could ensure a steady income for the farmers. In short, sericulture could fit perfectly into the homestead farming, which is very typical to the Kerala Agricultural scenario.

❑❑❑

Chapter 37

Spawn Production and Cultivation of Oyster and Milky Mushrooms

Dr. Geetha D. (Professor) and **Sreenath P.** (Research Scholar)

Kerala Agricultural University, College of Agriculture Vellayani, Thiruvananthapuram-695 522

Mushrooms with their flavour, texture, nutritional value, very high productivity per unit area and time, ability to grow on a variety of residual agricultural wastes, have rightly been recognized by FAO as food contributing to the protein nutrition of the countries. Cereals as staple constituents of Indian diet are deficient in two essential amino acids, namely lysine and tryptophan. Mushrooms, which are rich in lysine and tryptophan, can effectively supplement the cereals in terms of protein. It is also important to realize that they contain vitamins not occurring in green plants, such as B1, B2, B6, B12, D1, D2, H and Pantothenic acids. Their low sugar, low calorie and high fibrous nature are the added advantages.

Edible Mushrooms Cultivated in Kerala

Pleurotus/Oyster mushroom

Milky mushroom

Cultivation of Oyster Mushroom (Pleurotus)

About 70 different types of mushrooms are reported under the genus Pleurotus. Most of these are related to each other. Adequate temperature setting is a must for pleurotus cultivation. Pleurotus florida, P. sajor caju, P. eous, P. citrinopileatus, P. opuntia, P. fossulatus are the commonly cultivated Oyster mushrooms in Kerala.

Medium for Cultivation

Paddy straw, Rubber sawdust, Banana waste, etc. are the most commonly used media for mushroom cultivation. Agricultural wastes may be used too. Always use a high quality medium, which is clean and germ free.

Paddy straw: Use dry, clean, hard, germ-free, fresh, golden coloured straw to get a high yield.

Rubber saw dust: Use fresh and pure rubber sawdust. Do not mix it with any other sawdust. The final harvest may take upto 25-35 days but the mushrooms will be large and heavy, when rubber sawdust is used as medium.

Plantain trunks: The trunks are first cut into small pieces and sundried well. Infection is more in this type of media. Anyway, plenty of large mushrooms may be harvested from this type of media.

Sterilization

Sterilization is an important part of mushroom cultivation. There are three methods for sterilization:

Boiling

Keep the paddy straw soaked in water for about 6-8 hours. Clean it and drain the excess water. Add water and boil for 30-45 minutes. Then spread it on a clean floor and allow it to cool.

Steaming

Keep the paddy straw soaked in water for 6-8 hours. Clean and drain. Steam the straw for 45 minutes. A boiler may be used for this purpose.

Chemical Treatment / Pasteurization

This is the easiest method of sterilization. Chemicals are used to remove germs from the medium.The commonly used chemicals are Formalin (formaldehyde-40%) and Bavistin (Carbendazim). Mix 50 ml of Formalin and 7.5 g of Bavistin in 100 L water. Stir the mixture well. Soak the straw in the liquid for about 16-18 hours. Drain water so as to retain 50% moisture and then use the straw for cultivation. Store these in plastic buckets and cover tightly. Only then sterilization will take place effectively with the aid of chemical reaction. Spread the drained straw on a clean floor. Make sure that at least 50-60% water content is lost. Presence of water affects the growth of the fungus adversely and increases chances of decay. Plantain trunks need to be soaked in water for long hours. These may be boiled directly in water or even steamed. Rubber sawdust are to be kept in sacks and then need to be soaked in water. Boiling as well as steaming helps in sterilizing them. For chemical treatment, the sacks are to be soaked in chemical solution for about 18 hours. Drain and dry the powder. Reduce the water content to 50-60% and use.

Method of Cultivation

Oyster mushrooms are grown in polybags. Polybags with 150 gauge thickness and 60×30 cm dimension are ideal. A single spawn (300g) can be used for 2-3 mushroom beds. Fold the bag in to four and make tiny perforations using a needle.

Now, stack the straw in to layers in the bag. Each layer should be 6-7 cm thick. Sprinkle 25 g of spawn along the sides of the straw. Add another layer of straw as above, followed by spawn. Repeat until the cover gets filled. There will be about 5-6 layers in one bag. The top layer is spawned fully. Press down slightly and tie the bag. It will have a cylindrical shape. This is a typical mushroom bed. Plantain trunks or rubber sawdust may be used instead of straw. These mushroom beds are to be kept in a well aerated, cold dark room.

Mushroom Culivation – Different Stages

Spawn run

The time taken for the growth of the mushroom fungus mycelia in the medium is known as spawn run. Pleurotus florida takes about 15 days for spawn run. P. eous, Ananthan, P. opuntia, etc. take about 8-10 days to complete spawn run.

Formation of Mushroom Pinheads

The mushroom beds that have completed the spawn run are white in colour and the mycelia grows into a hard web-like structure. The beds that are not white in colour and have a soft mycelia growth indicate a poor growth of fungus. They can be kept under observation for 2-3 days. Infected beds yield poor growth and if there is no progress, destroy the beds. Make tiny slits of about 1 cm long on the beds. Keep the beds in a well-aerated, cold room with light. Temperature should be set between 26 and 28 degrees with a relative humidity of 90-100. To ensure proper aeration, water may be sprinkled on the walls of the room or lay a sheet on the floor and sprinkle water on it.

Harvesting

Mushroom pinheads appear in 3-4 days. Once the pinheads start to appear, you may avoid sprinkling water on the beds. But continue to sprinkle water on the walls or sheets. In a couple of days, the mushrooms grow fully and the first flush can be harvested. After the first harvest, clean the remains of the mushrooms present and sterilize the whole room. Sprinkle a solution of 1% bleaching powder solution in the mushroom room.

Cultivation of Milky Mushroom

Calocybe indica commonly known as milky mushrooms are the ideal ones for cultivation in Kerala. These are white in colour and are larger in size. They have a longer shelf life. Milky mushrooms are rich in fiber. The natural habitat of milky mushrooms in base of coconut or areas rich in cowdung and agricultural wastes. Mushrooms weighing about one and a half kg may be obtained naturally.

Temperature required for milky mushroom cultivation is between 28 and 30 p C and appropriate humidity is to be maintained as well. No strict climatic condition is required for growing milky mushrooms. Mushroom bed can be made using mearly 300-500 g of straw. Such a small bed can yield about 400-500 g of mushrooms. The mushroom look like button mushrooms but have a longer shelf life and can remain fresh up to a week. Thin quality makes milky mushrooms unique.

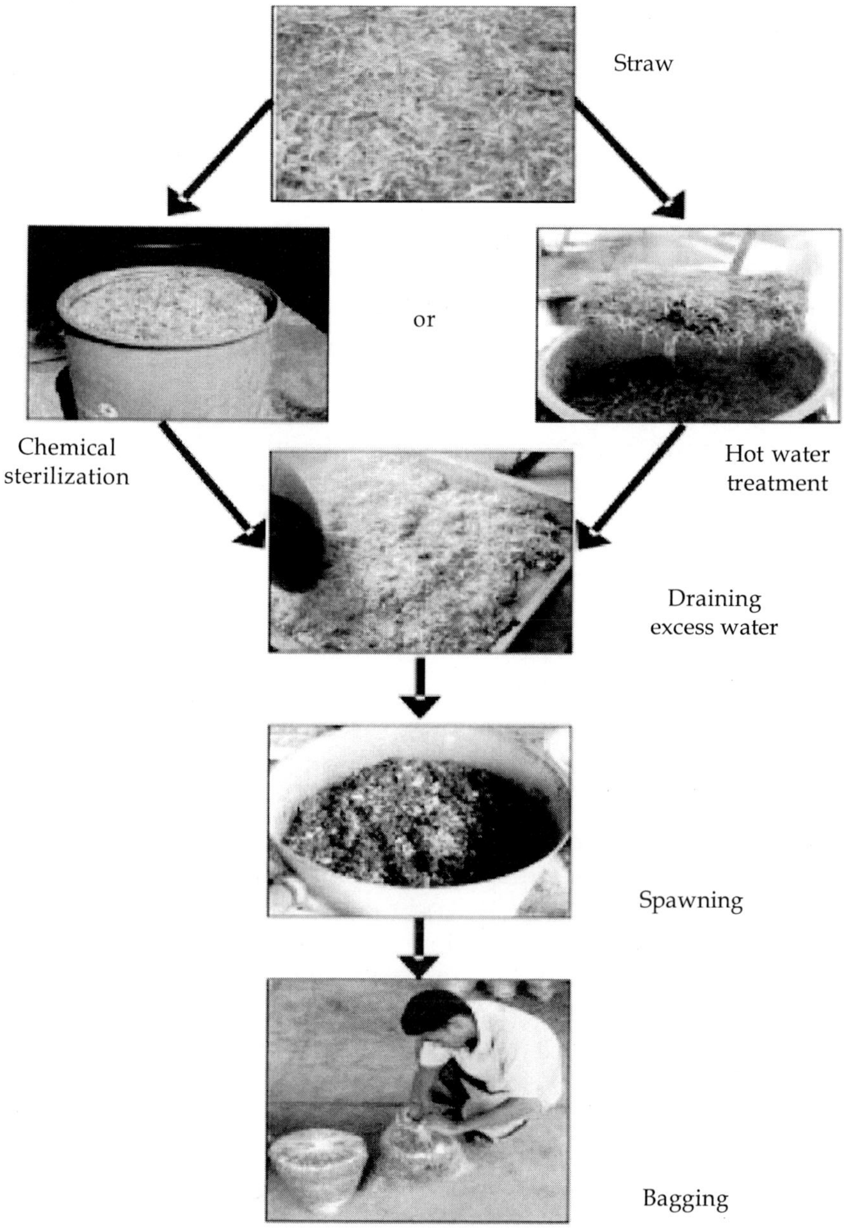

Different steps in oyster mushroom cultivation

Method of Cultivation

Materials required: Polythene covers (45×30 cm), straw/rubber saw dust, mushroom spawn.

Sterilize the straw or the rubber sawdust by boiling or steaming them. Allow only 50-60% moisture content to remain and drain the rest. Stack the straw into layers in a perforated polybag. A single layer should be 7-8 cm thick. Now sprinkle about 25g spawn all over the stacked straw. Stack the straw and spawn alternately. Repeat three times. The topmost layer should be covered with spawn. Tie the bag. A single mushroom spawn can be used for at least four beds made of three layers. The mushroom room should maintain a temperature of 30-32 p C with proper ventilation. Cleanliness should not be comprised. Mycelial growth happens in 15-20 days and the bed turns pure white. Then open the bag and do casing on top of each bed.

Casing

The ingredients required for casing media are a mixture of sand, soil and vermicompost or coir pith compost, calcium carbonate (2%) and water.

To prepare the casing medium, mix sand, soil and vermicompost in equal ratios. Add 20 g calcium carbonate (for every kg of the mixture), mix with water so as to retain 60% moisture content. Coir pith is also a good substitute for vermicompost. Unite the beds which have completed spawn run. Spread the casing medium on top of the bed in one inch thickness. Press down and keep them in the mushroom rooms. Casing boosts the growth of mushrooms. Also, maintain essential moisture content. Casing also helps in the uniform growth of mushrooms. After 8-10 days of casing, a white growth may be seen spreading through out the medium. And in 2-3 days mushroom pinheads start to appear and ready to harvest milky mushrooms are obtained in 5-7 days. A single mushroom bed yields 250- 500 g of mushrooms. Remove the mycelial bits and pinhead rudiments from casing surface after the first harvest and sprinkle with water and press down. Second harvest may be done in 10-15 days. A total of 500-600g mushrooms can be obtained from a single bed through 2 or 3 harvests. Milky mushrooms are quite heavy and a single mushroom weighs between 50g and 500g. Pest and disease incidence in milky mushrooms is comparatively less. However, inky cap mushrooms, green mold, etc. that grows on the casing media may pose serious threats. Snails that feed on mushrooms too are menaces. Sprinkling salts helps in reducing snail attack.

Casing Mixture

Layering of Casing Mixture

Fungal Growth of Casing

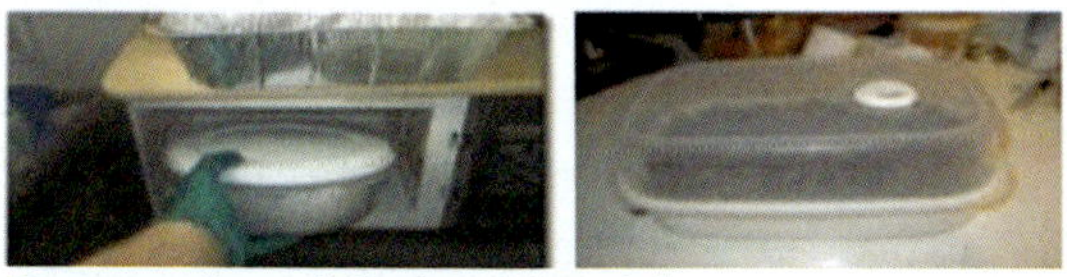

Preservation of Casing Material

Green Mould Growth on Casing

Pre-Treatment of Formalin

Healthy Mushroom Growth

Preparation of casing mixture

Pest and Disease Management in Mushrooms

- Pest control should be given prime importance while constructing a mushroom house. Use a 36-40 wire mesh to cover the windows and doors, which prevent the entry of pests to a great extent. Use a pesticide to clean the room at least once a week. Neem oil-garlic mixture or Nuvan or formalin can be used for this.
- Sterilization of straw should be done properly.
- Check the mushroom beds daily for any infection. If found, remove the beds.

- Always use good quality spawns.
- After crop, disinfect the room using formalin (0.005%) and keep the room closed for two days. Then allow proper aeration and start the next cultivation.

This breaks the lifecycle of the pests and thereby prevents them from attacking the mushroom beds.

Government Schemes and Assistance to Support Mushroom Cultivation

- National Horticultural Board (NHB) is giving credit linked back-ended subsidy@ 20% of the total project cost limited to Rs. 25 lakh per project to start commercial mushroom cultivation.
- Mushroom Development Foundation (MDF), a non-profit organization in Assam, supports small scale farmers to cultivate, produce and market mushroom as a livelihood opportunity.
- There is a central scheme by ministry of food processing, Government of India to train, disseminate information, provide technical and financial assistance for preparation of cultures/ mushroom spawn cultivation, harvesting, storage, processing, packaging, and generating income with special emphasis on women.
- Financial assistance and soft loans are available from National Co-operative Development Corporation (NCDC), and NHB.
- Directorate of Mushroom Research (DMR), Solan, Himachal Pradesh is giving training on mushroom cultivation.
- Under KSCSTE, WGDP, RKVY Projects, Mushroom units were established in different states and they provide training programme on mushroom cultivation technology.
- In Kerala, Kerala Agricultural University is giving training on mushroom cultivation and mushroom beds are provided including spawns.

Chapter 38

Value Added Products of Coconut

Dr. P.R. Geetha Lakshmi (Asst. Professor)

Department of Post Harvest Technology, Kerala Agricultural University, College of Agriculture Vellayani, Thiruvananthapuram-695 522

Coconut is known as 'Kalpavriksha' as almost all parts of the tree is useful. It is used as food, oilseed, fibre crop and coconut leaves, trunk, inflorescence and shell are also utilised. Even though we are third in world production of coconut behind Indonesia and Philippines, hardly 2% of production is used for processing to value added products. There exist a huge demand for coconut value added products in both, domestic and export market. Coconut value addition can play a pivotal role in foreign exchange earnings by offering a wide range of quality products. We are exporting as well as importing coconut products and the export is increasing for the last years. Majority of the entrepreneurs are unaware how to explore the opportunity by marketing a range of value added coconut products with good quality using modern technologies of production, packaging and storage. Due to enormous scope and potential of coconut processing sector, new prospective entrepreneurs can venture to this sector to make our state as a coconut processing hub.

Several value added food products can be prepared from different parts of coconut *viz.*, wet coconut kernel, dry coconut kernel, mature coconut water, tender coconut water and coconut inflorescence sap.

A. Products from Wet Coconut Kernel

1. Desiccated Coconut Powder (DCP)

Desiccated coconut powder is a popular commercial product in both, domestic and international market. It is prepared from wet kernel of coconut with full maturity. The coconut is grated or disintegrated after removing the brown testa and desiccated/ dried to a moisture content of less than 3%. It is pure white in colour and used extensively in confectionaries, puddings, decorating cakes, pastries, filling chocolates, in other food preparations and also used as a substitute for fresh grated coconut. Now-a-days this product has demand in urban domestic markets of coconut producing states also. Depending on the size and form of desiccated coconut, it is graded as fine, medium and coarse. As per FSSAI standards (Food Safety and Standards Authority of India) DCP should be white in colour, free from foreign matter, insects, mould, and rodent contamination. The product shall have pleasant taste, flavour, free from rancidity and any evidence of fermentation. As per regulations, in DCP extraneous vegetable matter should not be more than 15 units/100 g, moisture content not more than 3%, total ash not more than 2.5%, oil content not less than 55 %, acidity of extracted fat pressed as lauric acid not more than 0.3%, sulphur dioxide (preservative) not more than 50 mg/kg.

The production technology consists of de-husking, de-shelling and removal of brown testa. This can be done manually using scrapers/ peelers or by machines. The coconut is cut open and washed thoroughly and grated using hammer/pin mills. Then it is steam blanched for 20 minutes and dried in hot air at 80-90^{0}C with occasional stirring till the moisture become less than 3%. After drying it is cooled, sieved, graded and packed in polythene/laminated pouches. The quality and yield of desiccated coconut powder depends on quality of coconuts used and on an average, 100 coconuts gives around 10 kg of desiccated coconut powder. The by-products, brown testa can be used for extracting oil, as it contains about 55% oil, coconut water for value added products and shell for charcoal/activated carbon/handicrafts.

Machinery needed for a DCP manufacturing unit are whole nut inspection conveyor, de-husking machine, de-shelling machine, brown skin (testa) removing machine, washing unit, disintegration unit, blanching unit, dryer with pre drying circuit and dust collection system, DC powder cooler, lump braker, vibro sieve, intermediate conveyors, packaging and labelling unit.

Desiccated coconut powder can be prepared as high fat and low fat types. Low fat desiccated coconut powder is prepared after extracting the coconut milk with a low fat content up to 38-40%. It is used for the preparation of coconut dietary fibre and coconut flour. This can also be prepared as by-product of coconut milk or virgin coconut oil production unit.

2. Coconut milk / Coconut cream

Coconut milk is obtained by extraction of fresh coconut gratings with or without addition of water. Coconut milk in concentrated form is known as coconut cream, from which coconut milk can be obtained by dilution with water. The production process involves selection of fully mature nut, de-husking, de-shelling, washing, blanching in hot water at 80°C for 10 minutes. Then grated using hammer mill and milk is extracted using screw press. The coconut milk obtained is filtered through vibrating screens and food additives (emulsifiers and stabilizers) were added. The coconut cream so obtained is pasteurised at 95°C for 10 minutes in plate heat exchanger. The cream is canned/ sterilised in retort. Coconut skimmed milk can be prepared by removing the cream from coconut milk. Sweetened condensed milk can also be prepared. The residue obtained after the extraction of milk can be dried and used in confectionary.

3. Coconut milk powder

Ready to use coconut milk powder is produced by spray drying of coconut milk. It can be used instead of fresh coconut milk for food preparations in households and food industry by dissolving in water. The technology is developed by Central Food Technological Research Institute, Mysore. Mature nuts are de-husked and de-shelled to separate the kernel. Washed kernels are blanched in hot water at 80°C for 10 minutes and grated. Coconut milk is extracted using screw press and filtered. It is then mixed with maltodextrin and other emulsifiers. Then it is fed to spray drier to obtain fine milk powder and packed in aluminium foil packets.

4. Coconut jam

Coconut jam with rich coconut flavour of good quality can be prepared from coconut milk or tender coconut pulp. The pulp/coconut milk is mixed with brown sugar, glucose, citric acid and other additives and cooked over a slow flame till it reaches the jam consistency. The product is hot filled in sterilised containers and sealed hermetically.

5. Coconut milk beverages

Coconut milk mixed with fruit juices adds a variety of taste to fruit beverage basket. 'Pinacolada', coconut milk mixed with pineapple juice is developed ny Nadukkara Agro Processing Company Ltd, Muvattupuzha. It is prepared by blending of coconut milk, pineapple juice and other ingredients, homogenised, sterilised, and aseptically packed. Several blended coconut fruit beverages are available now-a-days.

6. Virgin coconut oil (VCO)

Virgin coconut oil is known as 'mother of oils' and has high demand in functional food industry due to health benefits. It is the purest form of coconut oil, watery white in colour with mild to intense coconut aroma depending on the type of process of production. VCO can be prepared from fresh coconut meat or coconut milk or coconut milk residue. Fresh coconut meat is partially dried and VCO is extracted using screw press. VCO is prepared from coconut milk by boiling, which is the traditional method. VCO obtained through boiling method has intense aroma and light yellow in colour. The cold extraction method consists of fermentation method and centrifugation method, where there is no application of heat. The fermentation method is natural fermentation and the oil has some off-flavour. The VCO from centrifugation method yield good quality oil. Coconut milk is extracted from fully mature coconut and filtered. This is subjected to centrifugation, where oil and skim milk is separated. Vacuum dehydration of oil will remove excess moisture present in the oil. Then it is passed through the filter and packed in consumer packs.

7. Coconut chips and bites

Coconut chips are ready to eat snack without frying. Matured coconut kernel after removing the brown skin is blanched in boiling water mixed with 0.05% potassium metabisulphite for 15 minutes and drained. The kernel is sliced and are immersed in osmotic solution of sugar at 50^0B for one hour and drained. The chips are then dried in hot air dryer at 70-80^0C for 4-5 hours. The dried coconut chips are allowed to cool to room temperature and packed in laminated pouches and labelled. Instead of sweet chips, salted/flavoured coconut chips can also be prepared using suitable osmotic medium. CFTRI, Mysore has developed technology for coconut-based instant rice mix, chutney powder/mix and coconut bites.

In addition to these products, fresh coconut kernel, coconut syrup, sweetened condensed milk, yogurt, cheese, mayonnaise, honey, fermented beverage concentrate, margarine, biscuit, ladoo, noodles, coconut flour, coconut dietary fibre, etc. can also be prepared from wet coconut kernel.

B. Products from dry coconut kernel

1. Copra

Fully matured coconut kernel is dried to obtain copra. Two types of copra are produced: edible copra used for the edible purpose and milling copra for the extraction of coconut oil. Two types of edible copra are available: ball copra and cup copra, which has demand in North India.

2. Coconut oil

Coconut oil is extracted from well-dried copra with moisture content not more than 6%. The copra is cut into small chips in a copra cutter and then fed to steam jacketed kettles and cooked mildly at a temperature of 70°C for 30 minutes. It is then fed to oil expeller, where oil is extracted which is filtered and packed. The oilcake obtained as by-product can be marketed as cattlefeed.

C. Products from mature coconut water

1. Bottled coconut water

Bottled coconut water is gaining popularity. Coconut water from mature nut is collected, filtered and upgraded by adjusting sugar, acidity, pH, taste, etc., pasteurised and bottled. It can also be carbonated and marketed. The carbonated and non-carbonated beverages can be packed in aluminium and polyethylene laminated packages with a shelf life of six months at room temperature.

2. Nata-de-coco

Nata-de-coco is a gelatinous product prepared from mature coconut water by the action of cellulose bacteria, *Acetobacter aceti* subspecies *xylinium*. Sugar and acetic acid is added to the mature coconut water in particular proportion and the bacteria is inoculated and kept for 3 weeks. White coloured jelly forms at the top and is washed, cut into pieces and packed. It is used widely in desserts, fruit salads, fruit cocktails, etc.

3. Coconut jelly

Coconut jelly is prepared from tender coconut water and jelly stored in glass bottles recorded a shelf life of 6 months.

4. Coconut vinegar

Mature coconut water obtained as a by-product in copra making unit can be converted into vinegar using vinegar generators. The process involves fortification of coconut water with sugar and alcoholic fermentation by yeast and after which acetic acid fermentation is carried out by adding mother vinegar. It has wide scope in food industry, flavour industry and high demand for export in place of synthetic vinegar.

D. Products from tender coconut water

1. Packaged tender coconut water

Tender coconut water is a natural source of electrolytes, minerals, carbohydrates, amino acids and other nutrients with numerous medicinal properties. The technology for packaged tender coconut water in pouches/aluminium cans is developed by Defense Food Research Laboratory, Mysore with a shelf life of more than 6 months under normal condition.

2. Minimally processed tender nuts

The tender coconut is bulky to handle and husk contributes to about 2/3 of the total nut. Partial removal of husk to make handling, transportation and display easier is achieved through minimal processing of tender coconuts. The technology is developed by Kerala Agricultural University. The technology consists of partial removal of tender nut husk and dipping in a solution of 0.5% citric acid and 0.5% potassium metabisulphite for 3 minutes. The product can be stored at 5-7^{0}C for 24 days.

3. Snow ball tender nut

It is the tender coconut without husk, shell and testa, which is in ball shape and white in colour. This solid endosperm of tender coconut containing tender coconut water is a delicious dessert. Tender coconut of 8 months maturity is de-husked and a groove is made in the shell using a machine developed by CPCRI and 'snow ball' is scooped out using a scooping tool.

E. Products from coconut inflorescence

1. Neera

Neera is the sap collected from immature unopened coconut inflorescence. It is a highly nutritious health drink with low glycemic index. The sap collected is filtered, preservative added, pasteurised and packaged.

2. Coco sugar / Jaggery

Neera is filtered, boiled which solidify on cooling yield jaggery. The sugar is also prepared by clarifying neera and evaporating and concentrating in vacuum pans.

Other products from coconut inflorescence sap are syrup, treacle-concentrated form of sap, and fermented beverages like toddy, vodka, wine, vinegar, etc.

Technological Assistance

- Coconut Development Board, Kochi
- CFTRI, Mysore
- Kerala Agricultural University, Thrissur
- NIIST, Thiruvananthapuram
- DFRL, Mysore

Export promotion activities

Government of India has provided measures to boost India's coconut products export under Foreign Trade Policy, 2015-2020 under two schemes:

a) Merchandise Exports from India Scheme (MEIS): Under this scheme Government of India provides incentives for exporting notified goods/products to notified markets. The rate of benefit ranges from 2-5% of the realised Free on Board (FOB) value of exports.

b) Duty Drawback Scheme: It is a popular method of encouraging export. In this custom duties paid on the inputs or raw materials and service tax paid on the input services used in the manufacture of export goods are refunded.

Financial assistance

Financial assistance is given by Coconut Development Board, Kochi as subsidy.

Details of assistance, project proposal, machinery/plant details for each product are available at Coconut Development Board, Kochi. P.B. No.1021, Kera Bhavan, SRV Road Kochi-682 011, Ernakulam District, Kerala. Ph. 04842376265. Website: www.coconutboard.gov.in.

References

Coconut Development Board http://coconutboard.nic.in/

CPCRI, 2014. Value chain in coconut. Final report - National Agriculture Innovation Project. Central Plantation Crop Research Institute, Kasargod. http://naip.icar.gov.in/download/c2-204701-kasargod.pdf

Muralidharan K. and Jayashree A., 2012. Value addition, product diversification and byproduct utilization in coconut. https://www.coconutboard.in/does/article-valueaddition-director.pdf

❑❑❑

VI. Livestock Production Management

Chapter 39

Multi-Level Participatory Farming (MLPF): A Model for Reviving Agripreneurship in Kerala

Dr. Suraj, P.T.

Assistant Professor (Livestock Production Management) ,Kerala Veterinary and Animal Sciences University

ABSTRACT

Safe and delicious food is still considered to be the most "Unavoidable Luxury" in human life. Maximising the state agricultural production needs a state wide participation of people from all the strata of life. With the declining agricultural production and number of farmers, the large numbers of non-producers (consumers) have to think seriously about supporting and sharing the responsibility of agricultural production. Sharing the cost and responsibility of the farming by the consumers is one of the best alternatives to support the farmers. In addition to this, the active involvement of the interested parties like technologists, institutions and traders will ensure the sustainability of such a system. The multi-level participatory farming method envisages the development of group farming along with fixing of unit value with proper weightage for land productivity, labour, technological interventions, management and marketing. The total output and total value of produce can be shared among the investment groups based on the units of investment. Involvement of persons having knowledge and skill in marketing can enhance the economic return. Even the state departments of marketing can take part in marketing by purchasing units of investment in the system and ensure supply of commodities at a reasonable price. Registration with local self-government in the respective areas can give legal existence to the particular groups. However, an alternative and a more relevant one will be registration under the industry department, similar to that of Producer Company, if the groups really thrive well in their business model. "Multi-Level Participatory Farming" is one of the best alternatives to support the farmers as well as to protect the interest of the consumers. In addition to this, the active involvement of the interested parties like technologists, institutions and traders alongwith public control will ensure the sustainability of such a system.

Introduction

Safe and delicious food is still considered to be the most "Unavoidable Luxury" in human life. Maximising the state agricultural production needs a state-wide participation of people from all the strata of life. With the declining agricultural production and number of farmers, the large numbers of non-producers (consumers) have to think seriously about supporting and sharing the responsibility of agricultural production.

The issue of food security is in the centre stage, on account of failing monsoon and climate change. The developments in agriculture have occurred by the balanced mix of technology and science. Development of new processes for improving the productivity has achieved much in the technological side. However, the scientific development has mostly omitted the sociological aspects of agriculture, failing in development of new social methodologies for the support and development of agriculture for its sustainability.

Thumburmuzhy Model

Cattle Breeding Farm Thumburmuzhy has devised a participatory farming method under the revolving fund scheme for organic vegetable production with the following objectives:

1. To produce vegetable following organic practices
2. To utilize the land available for food production for local consumers
3. To provide employment opportunities for the labourers

Methodology

Kerala Veterinary and Animal Sciences University has provided revolving fund for the unit for income generation. Considering the social responsibility, the members in the farm has decided to initiate the production of organic vegetables through participatory method. The farm has identified plots that can be utilised for the vegetable cultivation. The officers and labourers formed smaller group and the identified areas were handed over to the groups for cultivation. The inputs were purchased and supplied from the revolving fund. Land preparation, planting and management were carried out by the members in the groups. The farm has fixed the sale price of the individual items. Thus, the University has invested on land and capital, the manpower supplied by the employees and labourers and the marketing organised by the KVASU unit. The harvested organic vegetables were sold to the employees and workers of the farm, local households and the excess to the local shops. For sharing the profit, the expenditure was deducted from the total receipt and then the profit was shared equally between the University and the participants. The 50% share of the profit pays return on capital and land utilization cost of the University. The other 50% share of the profit pays the manpower involved in the cultivation, harvesting and marketing operations.

Similar Available Models

There are many existing models for agricultural developments in India like co-operative model, SHGs, Producer Company and many more. However, co-operation between the heterogeneous interest groups has tried in a few one only. Community Shared/Supported Agriculture in Japan and Canada are the one where the consumers were supported the producers directly.

Teikei is a system of community – supported agriculture in Japan where consumers purchase food directly from farmers. Teikei is closely associated with small-scale, local organic farming and volunteer based non-profit partnership between producers and consumers. It is widely cited as the origin of community-supported agriculture around the world.

Community Shared Agriculture (CSA) of Canada is a socio-economic model of agriculture and food distribution. A CSA consists of a community of individuals who pledge support to a farm operation where the growers and consumers share the risks and benefits of food production. CSAs usually consist of a system of weekly delivery or pick-up of vegetables and fruits in a vegetable box, sometimes including dairy products and meat.

Replication and modified model for Kerala

A small-holder dominated country like India has to develop its own methodologies, sometimes out of the box solutions, to stabilize the price rise and to protect the interest of the farmers as well as consumers. Sharing the cost and responsibility of the farming by the consumers is one of the best alternatives to support the farmers. In addition to this, the active involvement of the interested parties like technologists, institutions and traders will ensure the sustainability of such a system. Hence, a multi-level participatory farming method is proposed for Kerala.

The multi-level participatory farming method envisages the development of group farming alongwith fixing of unit value with proper weightage for land productivity, labour, technological interventions, management and marketing. At the outset, the group will be formed based on the agricultural produce/commodity. It can be a single commodity or a group. Then, based on the target output and inputs required, the proportion of units in each category like land, labour, cash investment, technological and marketing intervention, etc. can be fixed.

Pooling the productivity of small holdings without hindering the interest of its owners is the most challenging factor in developing this methodology. Fixing a unit value based on the size and giving a weightage based on the productivity will be a suitable method. Hence, a large area can be covered under a production group, which can substantially reduce the production cost and increase the quantity of production.

Investment Units

Total investment can be calculated based on the land, labour, capital and technology. The labour units can be calculated as man hours. This will provide the labourers and resource- poor-farmers to find a share in the system. The fixing of labour in terms of working hours can attract people from all sections of the society, both, as an investment and as a way to physical exercise or leisure.

Technological interventions and regular monitoring by the expert will give a boost to the system. A nominal unit value of maximum 1% can be allotted for the technological and management attribute. The National Agricultural Research System (NARS) including the Agricultural/Veterinary/Fisheries Universities also can utilise this facility for generating their own income as well as to ensure job opportunities to their outgoing students, especially the certificate and diploma holders. Technical groups including academic institutions and professional bodies can involve in this manner. Since India (states also) is having one of the largest networks of agricultural institutions, this network can act as a back bone to the system. The consumers can invest in the system by purchase of units by cash investment and can claim the preference and privilege in the regular supply of produce for their household use. A specific opportunity may be available for the hotlers and supply chain managers in investing on their raw materials of superior grades.

The total output and total value of produce can be shared among the investment groups based on the units of investment. Involvement of persons having knowledge and skill in marketing can enhance the economic return. Even the state departments of marketing can take part in marketing by purchasing units of investment in the system and ensure supply of commodities at a reasonable price.

Organisation of Production Groups

The present agricultural situation demands different production groups. The demand for cereal crops is very high. Hence, one of the prominent or preferential areas should be cereal group. Mechanisation should be a thrust area and the specific groups involved in this should be equipped with trained manpower and equipments.

The groups involved in livestock production should ensure that the crop residues are utilized to a maximum to convert that to valuable animal produce. As a sustainable agricultural production, system depends on farm yard manure, the livestock group should ensure that manure management is followed effectively. The potential of wastelands and underutilised land should be exploited to the maximum for fodder, feed ingredients and agro-forestry development for an overall sustainable development in agriculture.

In livestock production groups, the necessity for creation of fixed assets in the form of animal houses/shelters is inevitable. For application of precision farming and modern technologies in agriculture also reasonably long-term commitment of land is inevitable. Hence, a fixed term commitment of land for building the structures should be ensured. In addition to that make shift type of structures are to bc developed with minimum investment on the basement, for facilitating the shifting of the structures in necessary occasions. Responsible groups formed out of socially committed individuals will also promote environment friendly farming.

Public Control

Registration with local self-government in the respective areas can give legal existence to the particular groups. However, an alternative and a more relevant one will be registration under the industry department as a Limited Liability Partnership Company. In areas of labour scarcity, production groups can also be linked with MGNREGS, providing investment option in terms of labour to the poor in the system. According to the new Corporate Social Responsibility (CSR) policy of Govt. of India, it is mandatory for the companies to utilize 2% of the net profit for CSR activities. Since agriculture is a priority sector, the companies may be allowed to utilize the amount in "Multi-Level Participatory Farming Companies" for triggering its initial growth. In addition to this, there is enough scope of linking the system with the farmer's interest group as designed in Agricultural Technology Management Agency (ATMA), which are autonomous bodies of extension planning and reform of each district. Hence, the government control and involvement in such a system can be ensured for a sustainable and efficient agricultural production system.

Conclusion

Sustainable agricultural development in the background of climate change is not only challenging but also inevitable. Ensuring food security to the growing population is the greatest challenge faced by the local as well as world leadership. The developments in agriculture have occurred by the balanced mix of technology and science. Development of new social methodologies for the support and development of agriculture is essential for its sustainability. A small-holder dominated country like India has to develop new social methodologies involving local leadership and resources to protect the interest of the farmers as well as consumers. "Multi Level Participatory Farming" is one of the best alternatives to support the farmers as well as to protect the interest of the consumers. In addition to this, the active involvement of the interested parties like technologists, institutions and traders alongwith public control will ensure the sustainability of such a system.

References

Carter, C., G. Rausser, and A. Smith, 2008. Causes of the Food Price Boom. Giannini Symposium on the Causes and Consequences of the Global Food Crisis. Oct. 10. Berkeley, CA.

Cline, W.R., 2007. Global Warming and Agriculture: Impact Estimates by Country. Washington DC: Peterson Institute.

Conant R.T. and Paustian K., 2002. Spatial variability of soil organic carbon in grasslands: implications for detecting change at different scales. Environmental Pollution 116: pp. 127-135.

Cotula, L., Vermeulen, S., Leonard, R. and Keeley, J., 2009. Land Grab or Development Opportunity? Agricultural Investment and International Land Deals in Africa, IIED/FAO/IFAD, London/ Rome.

FAO, 2007. Livestock's Long Shadow: environmental issues and options. FAO, Rome.

FAO, 2008. Food Outlook. November. Food and Agriculture Organization, Rome.

IFAD, 2009. (Draft) Comprehensive Report on IFAD's Response to Climate Change Through Support to Adaptation and Related Actions.

Lal, R. 2004. Soil carbon sequestration impacts on global climate change and food securities. Science. 304:1623–1627.

Rosenzweig, C. and M. L. Parry, 1996. Potential impacts of climate change on world food supply: a summary of a recent international study. In: S. Ramphal and S. Sinding (eds.) Population Growth and Environmental Issues. Westport, CN: Greenwood Publishing Group.

Rowlinson, P., 2008. Adapting Livestock Production Systems to Climate Change –Temperate Zones. Livestock and Global Climate Change conference proceeding. May 2008, Tunisia.

Royal Society., 2001. The role of land carbon sinks in mitigating global climate change. Policy document 10/01.

□□□